Mammalogy: Adaptation, Diversity and Ecology

Mammalogy: Adaptation, Diversity and Ecology

Annabelle Webb

Callisto Reference

www.callistoreference.com

Callisto Reference,
118-35 Queens Blvd., Suite 400,
Forest Hills, NY 11375, USA

Visit us on the World Wide Web at:
www.callistoreference.com

ISBN: 978-1-64116-233-3 (Hardback)

Printed in China.

Cataloging-in-Publication Data

Mammalogy : adaptation, diversity and ecology / Annabelle Webb.
p. cm.
Includes bibliographical references and index.
ISBN 978-1-64116-233-3
1. Mammalogy. 2. Mammals--Adaptation. 3. Mammals--Ecology. I. Webb, Annabelle.
QL703 .M36 2019
599--dc23

Table of Contents

Preface

The vertebrate animals within the class Mammalia are called mammals. There are around 4,200 different species of mammals. They are characterized by the presence of a neocortex, three middle ear bones, hair and mammary glands. Mammals are adapted for life on land, in the sea, in trees, underground or in the air. They are intelligent animals with large brains, self-awareness, vocalization and organization. They communicate through the processes of scent-marking, singing, ultrasound production, echolocation, etc. Mammals can be solitary and territorial, but can also organize themselves into hierarchies, harems and fission-fusion societies. The studies of all mammals, their ecology, diversity, adaptation and management are under the field of mammalogy. This book is a compilation of chapters that discuss the most vital concepts in the field of mammalogy. It provides comprehensive insights into this vast area of study. Coherent flow of topics, student-friendly language and extensive use of examples make this book an invaluable source of knowledge.

To facilitate a deeper understanding of the contents of this book a short introduction of every chapter is written below:

Chapter 1- Mammals are vertebrates within the class Mammalia, which possess mammary glands, hair, a neocortex and three middle ear bones. The study of mammals is undertaken by the science of mammalogy. This chapter has been carefully written to provide an overview of mammals, their classification, evolution and reproduction.

Chapter 2- Some large carnivorous mammals are tiger, lion, leopard, jaguar, cheetah, puma, bear, etc. The topics elaborated in this chapter provide a detailed description of such large carnivores for a holistic understanding of the subject.

Chapter 3- Ungulates are large mammals, such as rhinoceroses, horses, pigs, giraffes, cattle, etc. These can be classified as odd-toed and even-toed ungulates. This chapter closely discusses about ungulates and their different types to provide an extensive understanding of ungulate mammals.

Chapter 4- Aquatic mammals are the mammals that dwell in water bodies. Examples of such mammals are the blue whale, Amazon river dolphin, elephant seal, walrus, etc. All the significant kinds of aquatic mammals inhabiting the seas, oceans and freshwater bodies of the world have been carefully analyzed in this chapter.

Chapter 5- A primate is a mammal that belongs to the order Primates. This chapter discusses in extensive detail about primates and in particular, the eastern gorilla, which is the largest species within the order.

I owe the completion of this book to the never-ending support of my family, who supported me throughout the project.

Annabelle Webb

Chapter 1

Mammals: Classification, Evolution and Reproduction

Mammals are vertebrates within the class Mammalia, which possess mammary glands, hair, a neocortex and three middle ear bones. The study of mammals is undertaken by the science of mammalogy. This chapter has been carefully written to provide an overview of mammals, their classification, evolution and reproduction.

Mammalogy

Mammalogy is the scientific study of mammals. Interest in nonhuman mammals dates far back in prehistory, and the modern science of mammalogy has its broad foundation in the knowledge of mammals possessed by primitive peoples. The ancient Greeks were among the first peoples to write systematically on mammalian natural history, and they knew many mammals not native to Greece; Aristotle recognized that whales and dolphins (cetaceans), although fishlike in form, are mammals allied to terrestrial furbearers. Until the late 18th century, much scientific work on mammals was devoted to taxonomy or to the practical matters of animal husbandry. The scientific explorations of the 19th century resulted in large collections of specimens from virtually all parts of the world. Most of the world's mammal species are believed to be known to science (with the possible exception of a good many rodent and bat species), but the biology of many species is totally unknown. Modern mammalogy is a multidisciplinary field, encompassing specialists in anatomy, paleontology, ecology, behavior, and many other areas.

Mammalian taxonomy traditionally relied largely on museum collections of preserved skins (with their skulls), but by the second half of the 20th century, additional information was being gained from other studies—e.g., behavior, genetics, and biochemistry. In both laboratory and field research, new techniques and instruments have opened avenues of research that had previously been difficult or impossible. The self-contained underwater breathing apparatus (scuba), for example, has been important in many aspects of marine mammalogy. Telemetry, the use of minute radio transmitters to convey information to the researcher from a free-living animal, has been a particularly useful tool, allowing the tracking of the animal in its natural state and the monitoring of physiological information. Video technology also has come into use, while in the laboratory a rapidly increasing array of molecular techniques have changed the way mammalogists determine evolutionary relationships (phylogeny) as well.

Mammal

Mammals are warm-blooded animals. This means they lots of heat inside their body, which keeps them warm. In order to create heat, mammals need plenty of fuel in the form of food. Most mammals, like this giant panda, have a coat of hairs to trap warm air near the skin. Other mammals, such as dolphins, spend all their time in water, where a furry coat wouldn't work. Instead, their skin has a thick layer of fat that traps heat. All baby mammals are nourished by milk produced by their mother. Because of this, baby mammals are dependent on their parents in the earliest years of their life.

All mammals share at least three characteristics not found in other animals: three middle ear bones, hair, and the production of milk by modified sweat glands called mammary glands. Hair has several functions, including insulation, color patterning, and aiding in the sense of touch. All female mammals produce milk from their mammary glands in order to nourish newborn offspring. Thus, female mammals invest a great deal of energy caring for each of their offspring, which has important ramifications in mammalian evolution, ecology, and behavior.

Although mammals share several features, Mammalia contains a vast diversity of forms. The smallest mammals are found among the shrews and bats, and can weigh as little as 3 grams. The largest mammal, and indeed the largest animal to ever inhabit the planet, is the blue whale, which can weigh 160 metric tons (160,000 kg). Thus, there is a 53 million-fold difference in mass between the largest and smallest mammals! Mammals have evolved to exploit a large variety of ecological niches and life history strategies and, in concert, have evolved numerous adaptations to take advantage of different lifestyles.

Habitat, Physiological Characteristics and Behavior

Different species of mammals have evolved to live in nearly all terrestrial and aquatic habitats on Earth. Mammals inhabit every terrestrial biome, from deserts to tropical rainforests to polar icecaps. Many species are arboreal, spending most or all of their time in the forest canopy. One group (bats) has even evolved powered flight, only the third time that this ability has evolved in vertebrates (the other two groups being birds and extinct Pterosaurs).

Many mammals are partially aquatic, living near lakes, streams, or the coastlines of oceans (e.g., seals, sea lions, walruses, otters, muskrats, and many others). Whales and dolphins (Cetacea) are fully aquatic, and can be found in all oceans of the world, and some rivers. Whales can be found in polar, temperate, and tropical waters, both near shore and in the open ocean, and from the water's surface to depths of over 1 kilometer.

All mammals have hair at some point during their development, and most have hair their entire lives. Adults of some species lose most or all of their hair but, even in mammals like whales and dolphins, hair is present at least during some phase of ontogeny. Mammalian hair, made of a protein called keratin, serves at least four functions. First, it slows the exchange of heat with the environment (insulation). Second, specialized hairs (whiskers or "vibrissae") have a sensory function, letting an animal know when it is in contact with an object in its environment. Vibrissae are often richly innervated and well supplied with muscles that control their position. Third, hair affects appearance through its color and pattern, helping to camouflage predators or prey as well as signal to predators a defensive mechanism (for example, the conspicuous color pattern of a skunk is a warning to predators). Hair also communicates social information (for example, threats, such as the erect hair on the back of a wolf; sex, such as the different colors of male and female capuchin monkeys; or the presence of danger, such as the white underside of the tail of a white-tailed deer).

Fourth, hair provides some protection, simply by providing an additional protective layer (against abrasion or sunburn, for example) or by taking on the form of dangerous spines that deter predators (porcupines, spiny rats, others).

Mammals are typically characterized by their highly differentiated teeth. Teeth are replaced just once during a mammal's life (a condition called diphyodonty). Other characteristics found in most mammals include: a lower jaw made up of a single bone, the dentary; four-chambered hearts; a secondary palate separating air and food passages in the mouth; a muscular diaphragm separating thoracic and abdominal cavities; a highly developed brain; endothermy and homeothermy; separate sexes, with the sex of an embryo being determined by the presence of a Y or 2 X chromosomes; and internal fertilization.

Development

There are three major groups of mammals, each united by a major feature of embryonic development. Monotremes (Prototheria) lay eggs, which is the most primitive reproductive condition in mammals. Marsupials (Metatheria) give birth to highly altricial young after a very short gestation period (8 to 43 days). The young are born at a relatively early stage of morphological development. They attach to the mother's nipple and spend a proportionally greater amount of time nursing as they develop. Gestation lasts much longer in placental mammals (Eutheria). During gestation, eutherian young interact with their mother through a placenta, a complex organ that connects the embryo with the uterus. Once born, all mammals are dependent upon their mothers for milk. Aside from these few generalities, mammals exhibit a diversity of developmental and life history patterns that vary among species and larger taxonomic groups.

Reproduction

Most mammalian species are either polygynous (one male mates with multiple females) or promiscuous (both males and females have multiple mates in a given reproductive season). Because females spend a long period of time in gestation and lactation, it is often the case that males can produce many more offspring in a mating season than can females. Thus, the most common mammalian mating system is polygyny, with relatively few males fertilizing multiple females and many males fertilizing none. This causes intense competition between males in many species, and the potential for females to be discriminating in choice of sire for offspring, leading to complex behaviors and morphologies associated with reproduction. Many mammal groups are marked by sexual dimorphism as a result of selection for males that can better compete for access to females.

About 3 percent of mammalian species are monogamous, males mating with a single female each season. In these cases, males provide at least some care to their offspring. Often, mating systems may vary within species depending upon local environmental conditions. For example, when resources are low, males may mate with a single female and provide care for the young. When resources are abundant, the mother may be able to care for young on her own and males will attempt to sire offspring with multiple females.

Other mating systems, such as polyandry, can be found among mammals. Some species (e.g. common marmosets and African lions) display cooperative breeding, in which groups of females, and sometimes males, share the care of young from one or more females. Naked mole rats have a unique mating system among mammals. Like social insects (Hymenoptera and Isoptera), naked mole rats are eusocial, with a queen female mating with several males and bearing all of the young in the colony. Other colony members assist in the care of her offspring and do not reproduce themselves.

Mammals range from many altricial young in each bout of reproduction (rodents and insectivores) to those species that give birth to one or a few precocial young. The former tend toward high early mortality and short lifespans, while the latter invest energy in a few offspring that develop into efficient competitors, living longer in generally stable environments (Cetaceans, primates and artiodactyls). Most mammals make use of a den or nest for the protection of their young. Some mammals, however, are born well developed and able to locomote on their own soon after birth. Most notable in this regard are artiodactyls such as wildebeest or giraffes. Cetacean young must also swim on their own shortly after birth.

Generally, smaller mammals live short lives and larger mammals live longer lives. Bats (Chiroptera) are an exception to this pattern, being relatively small mammals that can live for one or more decades in natural conditions, considerably longer than natural lifespans of significantly larger mammals. Mammalian lifespans range from one year or less to 70 or more years in the wild. Bowhead whales may live more than 200 years.

Behavior

There are mammal species that exhibit nearly every type of lifestyle, including fossorial, aquatic, terrestrial, and arboreal lifestyles. Locomotion styles are also diverse: mammals may swim, run, bound, fly, glide, burrow, or climb as a means of moving throughout their environment.

Social behavior and activity patterns vary considerably as well. Some mammals live in groups of tens, hundreds, thousands or more individuals. Other mammals are generally solitary except when mating or raising young. Mammals may be nocturnal, diurnal, or crepuscular.

Olfaction, hearing, tactile perception, and vision are all important sensory modalities in mammals. Olfaction plays a key role in many aspects of mammalian ecology, including foraging, mating and social communication. Mammalian hearing is well developed as well. In some species, it is the primary form of perception. Echolocation, the ability to perceive objects in the external environment by listening to echoes from sounds generated by an animal, has evolved in several groups, including microchiropteran bats (Chiroptera) and many toothed whales and dolphins (Odontoceti). Vision is well developed in a large number of mammals, although less important in many species that live underground or use echolocation.

Mammals can be carnivores (e.g., most species within Carnivora), herbivores (e.g., Perissodactyla, Artiodactyla), or omnivores (e.g., many primates). Mammals eat both invertebrates and vertebrates (including other mammals), plants (including fruit, nectar, foliage, wood, roots, and seeds) and fungi. Being endotherms, mammals require much more food than ectotherms of similar proportions. Thus, relatively few mammals can have a large impact on the populations of their food items.

Economic Impact on Humans

Mammals are a vital economic resource for humans. Many mammals have been domesticated to provide products such as meat and milk (e.g., cows and goats) or fiber (sheep and alpacas). Many mammals are kept as service animals or pets (e.g., dogs, cats, ferrets). Mammals are important for the ecotourism industry as well: people travel to zoos and throughout the world to see animals like elephants, lions, or whales. Bats often help control populations of crop pests. Norway rats and domestic mice are vitally important in medical and other scientific research, serving as models in human medicine and research.

On the other hand, some species have a detrimental impact on humans and the ecosystem. Many mammals that eat fruit, seeds, and other types of vegetation are crop pests. Carnivores are often a threat to livestock and even to human lives. Mammals common in urban or suburban areas have caused automobile accidents when straying into roads and have become household pests. A few species coexist exceptionally well with people, including some feral domesticated mammals (e.g., rats, house mice, pigs, cats, and dogs). As a result of intentional or unintentional introductions near human habitation, these animals have had considerable negative impacts on the local biota of many regions of the world, especially the endemic biota of oceanic islands.

Many mammals can transmit diseases to humans or livestock. The bubonic plague is perhaps the most well-known example, spread via fleas carried by rodents. Rabies, which can be transmitted among mammalian species, is also a significant threat to livestock and can kill humans as well.

Mammal Classification

Mammals are of different types and can be distinguished up into marine mammals, smaller mammals and larger mammals. Mammals belong to the class Mammalia.

Since mammals are of different types they are classified into three subclass based on their reproduction.

Basics for Classification

They include approximately 5000 living species and 15000 sub species and numerous fossil forms. Today the classification varies coming from various agencies and taxonomists. But the current classification is basically described on the basis of following characteristics:

1. Mode of caring for their young ones.
2. Nature of their dentitions present in the animal group.
3. Foot posture.
4. Nails, claws and hoofs.
5. Complexity of the nervous system.
6. Systemics.

The class Mammalia is classified into three subclasses (Eutheria, Metatheria and Prototheria), 28 Orders, 161 Families, 747 Genera and 4939 Species.

Eutheria

Eutheria ("true beasts") are a group of mammals consisting of placental mammals plus all extinct mammals that are more closely related to living placentals (such as humans) than to living marsupials (such as kangaroos). They are distinguished from non-eutherians by various features of the feet, ankles, jaws and teeth. One of the major differences between placental and non-placental eutherians is that placentals lack the epipubic bones, which are present in all other fossil and living mammals. The absence of these bones is probably an adaptation that helps them be faster on the run than their relatives.

Mildly recognizable eutherians broke away from the metatherian strain no later the Early Cretaceous in warm, probably moderately seasonal climates. Immediate ancestors were small, sharing many anatomical, physiological and reproductive features with small modern marsupials. Development of characteristically eutherian features involved interactions of body size and rates of metabolism (this means the rate of usage of energy).

Biological implications of this contrast are most pronounced at small body sizes. When resources are abundant, the relatively higher growth rates and earlier maturation of small eutherians (particularly those with high rates of metabolism) can lead to rapid population growth; among most marsupials, however, such high rates of reproduction are rarely possible. Also, only eutherians among mammals give birth to truly developed embryos and that has helped their spread across the globe.

Geographic Range

These mammals have truly mastered all sorts of environments. They range from the polar caps(polar bears) to the equator(gorilla) to the oceans(blue whale). Among other things Eutherians support vast orders that have the ability to completley modify their body temperature and heart beat. This helps them beat the odds in adverse climes like deserts and ice caps.

Behavior

These mammals are known to live in very harsh climates and have adapted to eat anything that they can, at an extremely fast pace. They are so diversified that there is no one behavior that can be truly called eutherian aside from giving birth to well defined and very active babies.

Communication

Eutherians are the best communicators among mammals. They live in solitary arrangements, in groups and even in vast herds millions strong(think wildebeest).

Predation

Eutherians have adapted to be niche based eaters. They generally eat everything and each other when time and opportunity arises.

Conservation

The most numerous eutherians are rodents and humans. They seem determined to have a population competition and the rodents are losing. We humans are eating and despoiling the planet for all the rest of the animals and so are the biggest threat to ourselves.

Orders within Eutheria

1. Superorder Afrotheria (Afrotherian mammals)
    - Order Afrosoricida (tenrecs and golden moles)
    - Order Hyracoidea (hyraxes)
    - Order Macroscelidea (elephant-shrews)
    - Order Proboscidea (elephants)
    - Order Sirenia (dugongs, manatees, and sea cows)
    - Order Tubulidentata (aardvark)
2. Superorder Euarchontoglires (Euarchontoglirean mammals)
    - Order Dermoptera (flying lemurs)
    - Order Lagomorpha (hares, pikas, and rabbits)
    - Order Primates (primates)
    - Order Rodentia (rodents)
    - Order Scandentia (treeshrews)
3. Superorder Laurasiatheria (Laurasiatherian mammals)
    - Order Carnivora (carnivores)
    - Order Cetartiodactyla (Whales, Deer, Pigs & Antelopes)
    - Order Chiroptera (bats)
4. Superorder Eulipotyphla (hedgehogs, shrews, and relatives)
    - Order Erinaceomorpha (gymnures and hedgehogs)
    - Order Soricomorpha (shrews, moles, desmans, and relatives)
    - Order Perissodactyla (horses, rhinoceroses, and tapirs)
    - Order Pholidota (pangolins)
5. Superorder Xenarthra (Xenarthrans)
    - Order Cingulata (armadillos)
    - Order Pilosa (sloths and anteaters)

Metatheria

These mammals, comprise 9 orders, 23 families and around 330 species (more may be found when people chart out the whole of Papua and New Guinea). Most of these species are concentrated in Australasia but some 100 odd species are also found in the Americas. These animals have usually been in competition with Eutherians and have given ground in areas where both had to fight for the same ecological niche.

In Australia, they have evolved to occupy all sorts of niches that the tradional moles, shrews, primates, carnivores and other Eutherians occupy in other parts of the world.

In times of yore, when the world was less cluttered with pushy primates, filled with self-love, the marsupials had come up with incredible types like large flesh-eating kangaroos, rhino sized 'roos and even a sabre-tooth marsupial cat.

Distinctive Characters

Distribution

Metatheria were widely distributed over many parts of the world as recently as the Miocene period. They are now confined to the Australian region (with the exception of New Zealand), to South America and (a few species) to North America. During Tertiary period they were widely distributed in South America.

The Australasian region, excluding New Zealand, contains more than four times as many marsupials as occur in the Americas, where their congregation is chiefly in the southern continent. Two families, Didelphyidae and Epanorthidae (Caenolestes) are found in the Neotropical region and the genus Phalanger extends into the island of Celebes, where it is represented by two peculiar species.

Geological history of the group is interesting. In Australia, which is now its chief home, no fossil remains are found before Pleistocene. In America, remains of both polyprotodonts and diprotodonts are found as far back as Eocene, and in the case of Epanorthidae (Genus Cimolestes, Telacodon and Batodon) in the upper Cretaceous.

Didelphyds are found in the Tertiaries of Europe, and there is a number of fossil forms in the Jurassic Formation of Europe and N. America which belong to the group polyprotodont. These continents were probably connected about 100 million years ago. By the time placentals appeared Australia and South America were no longer connected with the other continents and the marsupials became extinct in Europe and North America which were still connected.

South America remained isolated until the formation of the Panama bridge in the Pliocene and most of its marsupial fauna became extinct after this time. In North America also there were no marsupials between the middle Miocene and the Pleistocene, when opossums (Didelphis) appeared, probably from South America. Australian marsupials only became threatened by the advent of Man from Europe and the other placentals he brought with him.

However, large Macropus have thrived with the spread of agriculture. Marsupials diversified in Australia due to isolation, and differentiated structurally into numerous types, arboreal, fruit-eating, grazing, gnawing, digging, burrowing, ant-eating, insectivorous or carnivorous.

Habits and Habitat

Marsupials are terrestrial and carnivorous (e.g., native-cats, tasmanian Devil, marsupial wolf), terrestrial and herbivorous kangaroos; arboreal and insectivorous Didelphys, Marmosa, Chironectes, etc.; arboreal phalangers or opossums and semi-arboreal Phascogale and pouched rats and mice, and pouchless Jerboa-like marsupials and banded ant-eaters; burrowing desert marsupial mole and bandicoots; and flying and climbing phalangers.

Marsupials are herbivorous (e.g., kangaroos), insectivorous, carnivorous (thylacines and dasyures), and omnivorous, diurnal or nocturnal, warm-blooded, air-breathing, viviparous pouched animals.

External Features

Body is covered over with hairs. Pinna (external ear) is well developed, but absent in moles. Tail is generally long and prehensile, and an important organ of balance in kangaroo rats and jumping mice. Mammary glands are modified sebaceous glands and have elevated nipples.

Females usually have a marsupial pouch or marsupium but it is absent in Didelphys and the Dasyuridae (e.g., Dasyurus, Thylacinus, etc.). Marsupium encloses the nipples. The number of teats bears a relation to the number of young produced at a birth.

Exoskeleton

It includes ectodermal structures, the hairs derived from the Malpighian layer of epidermis and claws. Skin also contains sweat gland, sebaceous glands and scent glands.

Body Cavity

Typical muscular diaphragm divides the body cavity into thorax and abdominal cavities.

Endoskeleton

(i) Skull

Skull is dicondylic. Brain case small. Orbit and temporal fossa fully confluent and no postorbital bar. Bony palate is incomplete posteriorly. Jugal bone reaches back so as to participate in the formation of glenoid cavity for articulation of lower jaw. Lacrymal bone extends outside the orbit.

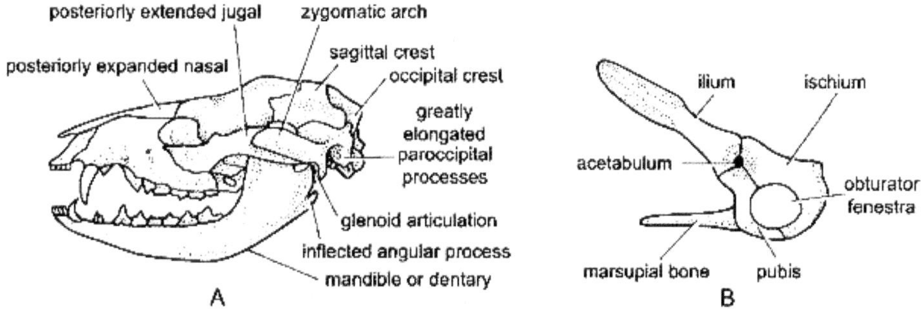

Opossum. (A) - Skull and mandible in lateral view; (B) - Pelvic girdle showing marsupial bone

Tympanic bulla is absent. Tympanic, periotic and squamosal remain separate. Alisphenoid assists in forming the bulla. In kangaroos alisphenolid extends backward to join the greatly elongated paroccipital process. The lower jaw, consisting of a single dentary, has inflected or inturned inner angle to which large lateral pterygoid muscle is attached.

(ii) Vertebral Column

The odontoid process fuses early with the axis. Cervical ribs with their respective vertebrae. Seventh cervical vertebra is pierced by a foramen for the vertebral artery. Presence of 13 rib-bearing thoracic vertebrae, as in placentals, and 7 lumbars as usual.

Sacrum has two vertebrae, but it may be reinforced by caudal vertebrae. Caudal region is short in koala and wombats, long in opossums, dasyures, phalangers and kangaroos. Chevron bones are generally present in caudal vertebrae, except in koala and wombats.

(iii) Girdles

In pectoral girdle, coracoid is reduced as in placentals, and scapula enlarged and provided with a spine. Clavicle is present except in bandicoots. There is no interclavicle. In pelvic girdle, epipubic bones project forward from the pubes in all. In Thylacinus these are represented by small cartilages.

(iv) Limbs

Forearm bones are separate and generally adapted for pronation and supination. Thumb is not opposable, but the two inner digits of the hand can frequently be opposed to the three outer in grasping. Hand with five clawed digits except in Chaeropus and the carpus is without centrale.

In hind limb five digits, fibula is free and can be rotated on the tibia. In some cases first digit can be used as a thumb. Hallux is frequently absent. Second and third digits are in most cases slender and united by the skin (syndactyly).

Digestive System

There is only one set of teeth functional throughout life with the exception of a milk molar which is replaced by the last premolar. Incisors are more numerous than in placentals. 3 premolars and 4 molars as against 4 P and 3 M in placentals. Dentition in opossum is I 3/4, C 1/1, P 3/3, M 4/4. Stomach is usually simple, but in kangaroos it is much elongated and sacculated. Caccum is usually present, absent in dasyures.

Respiratory, Circulatory and Excretory Systems

Respiratory, circulatory and excretory systems are typically mammalian. Respiration by lungs. Heart is without fossa ovalis. Auriculo-ventricular valves are membranous and attached to the papillary muscles by chordae tendineae.

Nervous System

Brain is relatively smaller. Olfactory bulbs large. Corpus callosum is absent and anterior commissure connects the two corpora striata (nerve band present in the lateral walls and floor of paracoels, lateral ventricle of cerebrum) is large. Cerebral hemisphere surface is convoluted. Cerebellum is small and simple. Cochlea of internal ear is spirally coiled.

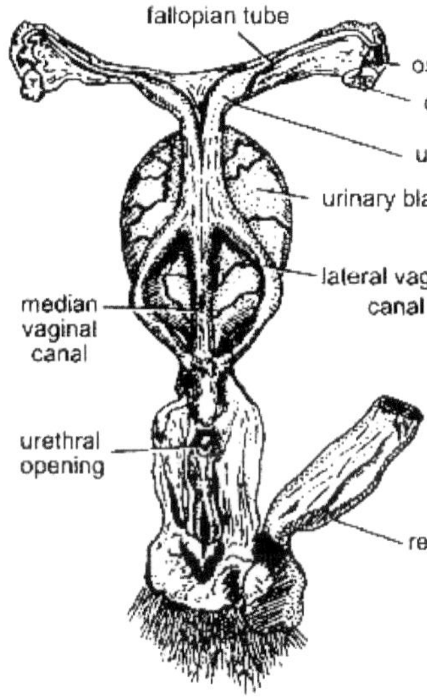

Female reproductive system of kangaroo.

Reproductive System

In male, there are no vesiculae seminales. Testes descend into scrotal sacs placed in front of the penis. Glans penis (tip of the penis) is bifid in several species.

In females, the Mullerian ducts are separate posteriorly and open separately into the long urinogenital sinus. They are differentiated into oviduct, uterus and vagina on each side (e.e., Didelphis). In other forms the anterior part of vagina gives of backwardly directed diverticula, the two meet in the midline as a sinus vaginalis and extend up to the front end of urinogenital sinus. This ends blindly until the young is about to be born. In some forms (e.g., Macropodidae) its hind end acquires an opening into the urinogenital sinus at parturition. Clitoris is also double.

The rectum and urinogenital sinus open together in a common cloaca. Cloaca is longer in female than the male.

Development

Females are viviparous. The egg is yolky and covered with albumen and a shell membrane. Cleavage is unequal. Placenta is of yolk sac type, vascular wall of yolk sac is in contact with the somewhat hypertrophied uterine wall. In Perameles (bandicoot) allantois develop a nutritive function to some extent.

Embryos are born very young as little as 8 days from conception in opossum. They are developed in the pouch attached to the teats. The young stays in the pouch for 8-9 months and the mother remains in lactationally induced anoestras.

Population Control

The males of marsupial mouse (Antechinus) die one week after mating. It is due to increased adrenocortical activity in the last weeks of life due to which they become more aggressive in nature. Females survive for two or three seasons.

Communication

As with mammals in general, vision, hearing, hearing and touch are all important to varying degrees among species of Metatheria. Communication can take many forms as well. Some marsupials communicate with sound signals, particularly during mating or territorial encounters. Many species have conspicuous color patterns that may convey information about sex or species identity.

Predation

Marsupials are rarely the top predators in the areas that they roam. Most of them are herbivores and are therefore open to the depredations of humans as well as alien species introduced into their environments.

Conservation

Many marsupials are threatened or endangered. The International Union for the Conservation of Nature and Natural Resources (IUCN) currently lists over 200 species (i.e., more than 2/3 of all marsupials) as being of some level of concern. Habitat destruction, overexploitation, and competition with exotic species and livestock have greatly reduced many populations. A number of species have gone extinct within the past two centuries as a direct result of human activity. The most famous among them is the Tasmanian tiger or thylacine.

Orders within Metatheria

1. Superorder Ameridelphia
    - Order Didelphimorphia (95 species)
    - Order Paucituberculata (6 species)
2. Superorder Australidelphia
    - Order Yalkaparidontia EXTINCT
    - Order Microbiotheria (1 species)
    - Order Dasyuromorphia (71 species)
    - Order Peramelemorphia (24 species)
    - Order Notoryctemorphia (2 species)
    - Order Diprotodontia (137 species)
    - Order Sparassodonta EXTINCT

Affinities of Marsupials

Marsupials show a mixture of primitive and advanced features:

1. Affinity with Prototheria:

- Presence of cloaca. The rectum and urinogenital sinus open together in a common cloaca.
- Presence of clavicles, epipubic bones in pelvic girdle and tympanic bone is ring-like.
- Tympanic bulla is absent. Tympanic is not united with the periotic.
- Brain is relatively smaller with large olfactory bulbs and anterior commissure. Corpus callosum is absent.
- True allantoic placenta is absent.
- Vesicula seminales are absent in males.

Differences with the Prototherians

- Metatherians have well develope pinna, absent in prototherians.
- Metatherians have well-developed marsupium.
- Vertebrae with epiphysis, Ribs bicephalous (monocephalous in prototherians). No interclavicle. Coracoids reduced as in placentals.
- Teeth present in the adults (only one set of teeth).
- Cochlea is spirally coiled, whereas it is partly coiled in Prototheria.
- Vivparous.
- Uterine gestation (No gestation in monotremes).
- Testes in the scrotal sacs, whereas in prototherians these are abdominal. Glans penis is bifid.

2. Affinity with Eutheria**:**

Metatherians have many advanced features similar to eutherians or higher placental mammals:

- Presence of hairs over the body.
- Presence of external ears (pinnae).
- Sebaceous mammary glands with teats.
- Coracoids reduced, interclavicle absent and ribs bicephalous.
- Teeth heterodont: incisors, canines, premolars and molars.

- Brain with 4 optic lobes. Cochlea spirally coiled.
- Testes in scrotal sacs. Presence of erectile penis in male.
- In female uterus and vagina are present in the oviduct.
- Viviparous. Ova microlecithal. Uterine gestation and placenta present.

Differences with the Eutherians

- Metatherians have restricted distribution, found only in Australia and Americas (North and South America).
- Presence of marsupial pouch.
- Number of incisors is more in both the jaws.
- Tympanic bone absent. Alisphenoid assists in forming the tympanic bulla. Penotic and squamosal remain separate. Jugal extending back and participates in forming the glenoid cavity for the articulation of lower jaw. Well-developed epipubic bones. Ventral posterior border of lower jaw is inwardly inflected.
- Corpus callosum is absent in the brain.
- In perameles, a true chorio-allantoic placenta is present but it is absorbed after the birth of the young. Others have yolk sac placenta.
- A pair of uteri and vaginae are present in females. Each vagina has a separate opening into the urinogenital canal.
- Ureters pass between the genital ducts, whereas in eutherians they pass outside the genital ducts.
- In male penis is sometimes bifid and behind the scrotal sacs.
- Gestation period is brief.

Systematic Position

It is obvious that the Metatheria are more advanced than the primitive, reptile-like, oviparous Prototheria. They are more closely related with the Eutheria, but do not belong to the same grade of evolution. Therefore, they are put under a separate infraclass Metatheria, while the rest of the higher and truly placental mammals are placed in the infraclass Eutheria and both combined in the subclass Theria.

Phylogenetic Consideration of Metatheria (Marsupials)

Comparative study of biological organisation of marsupials and placentals reveals that the marsupials appear to be 'second grade of mammals.' For this reason, it was the common practice for the zoologists to regard the marsupials as the transitional step in the evolution of mammals between the ancestral Jurassic mammals and the Coenozoic placentals.

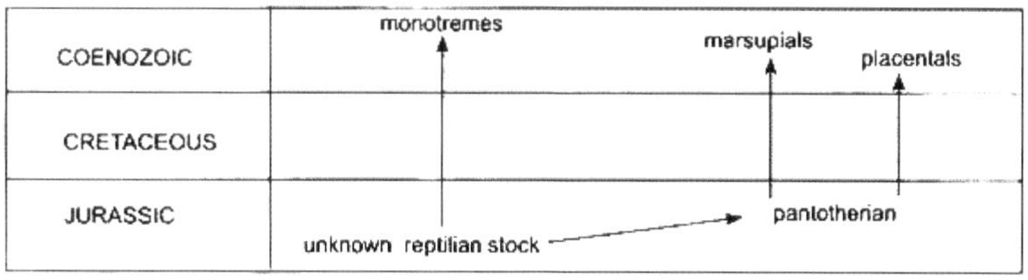

Showing phylogenetic relationship between three surviving mammalian groups.

But now it is believed that placental mammals and marsupials evolved independently from some common pantotherian ancestor in the late Jurassic period and both of them evolved side by side figure below will give an idea about the phylogenetic relationship of marsupials with the monotremes and placentals.

Prototheria

The subclass Prototheria contains the egg-laying mammals, which are the most ancestral forms in the class Mammalia. There are only five extant species grouped into two families and a single order, the Monotremata. Despite bearing fewer species than most mammalian genera, the prototherians are so unique among mammals that there is little question of them being.

Prototherians probably split from the lineage leading to other mammals sometime in the Mesozoic. They retain many characters of their therapsid ancestors (for example, a complex hip structure, laying of eggs rather than bearing live young, have one opening for excretion and urination etc.). The skulls of monotremes are almost birdlike in appearance, with a long beak and smooth external appearance. Monotremes have several important mammalian characteristics, however, including fur (but they lack whiskers), a four chambered heart, a single dentary bone, three middle ear bones, and the ability to produce milk(lactate).

Prototherians can be traced back to a lineage that began from reptile-like creatures approximately 190 million years ago. This group split from the ancestors of Eutherian and Metatherian a long time before those became placental and therefore more adapted to diverse climates and harsher environments. This group looks like a combination of reptiles, birds and "true mammals". This bizarre array of combinations makes them very interesting and also very hard to pin down to one class. It is only based on modern genetics we are able to put them firmly within mammals because they have the same Y-chromosome based determination of the sex of an individual.

Geographic Range

There are only a five species that still exist in this suborder and they are found in Australia and New Guinea. Old fossils found in Argentina indicate that once monotremes ranged a bit further that the boundaries of the Wallace Line.

Behavior

This group is mostly solitary with at least one species (Short-beaked Echidna) being territorial. Echidnas live on the ground and eat ants, termites and worms wher as platypuses spend most of their time in the water using their beaks to dig out worms from the stream beds. All extant species are exceptional diggers, using their powerful limbs to dig shelters. These burrows are a defence against predators as well as a great place to lay their leathery eggs. They can stay dormant for long periods of time if food sources are scarce.

Very little is known about the mating habits of prototherians except that they are found in pairs during breeding season. The young platypi use a milk tooth to break out of their shell after 12 days of incubation and the mother provide them milk and rears them for quite some time before they are able to fend for themselves.

Communication

Hearing, touch and smell are all important to some degree in prototherians. The platypus may even use electrical signals to find prey hidden underneath the stream or riverbed. A good sense of hearing is also important for Echidnas as they forage along the forest floor looking for prey. Very simple sounds that can emanate from beaks are sometime used but we do not known their functions well enough to discern whether they are mating calls or for protection of territory.

Predation

These are primitive creatures in that they have not had to evolve too much from the original ancestors so one is quite amazed to note that they have few protections against predation. Echidnas have spines, very much in the hedgehog fashion and functioning in a similar way. Platypi have a bit of venom that they can inject from the spur of their hind-feet. Although, venom is a rarity for mammals to possess, it is not much against sophisticated predators like man. These animals are not aggressive in nature and will not attack if left unmolested.

Conservation

Due to human intervention and the wholesale destruction of habitat at least three species of Echidna are at the brink of extinction. They are listed as critically endangered by the International Union for the Conservation of Nature (IUCN) redlist. Platypi were being hunted extensively for their fur but recent conservation efforts have helped steady their populations.

Order Monotremata

Monotremes probably split from the lineage leading to other mammals sometime in the Mesozoic. They are often placed in a separate subclass from other mammals, Prototheria. They retain many characters of their therapsid ancestors (for example, a complex pectoral girdle, laying of eggs rather than bearing live young, limbs oriented with humerus and femur held lateral to body, and a cloaca). The skulls of monotremes are almost birdlike in appearance, with a long rostrum and smooth external appearance. Modern monotremes lack teeth as adults; sutures are hard to see; the rostrum is elongate, beak-like, and covered by a leathery sheath; and lacrimal bones are absent. Monotremes have several important mammalian characters, however, including fur (but they lack vibrissae), a four chambered heart, a single dentary bone, three middle ear bones, and the ability to lactate.

Besides the absence of teeth, lacrimals, and obvious sutures, monotremes share a number of skeletal characteristics. On the skulls, the jugals are reduced or absent, the dentary is a slender bone with only a vestige of a coronoid process, the angle of the dentary is not inflected medially (unlike that of marsupials), auditory bullae are missing (part of the middle ear is enclosed by tympanic rings), and much of the wall of the braincase is made up by the petrosal rather than the alisphenoid (unlike all other modern mammals). Postcranially, the skeleton of monotremes is also unique among mammals. It is a fascinating mosaic of primitive characteristics inherited from therapsids but found in no other living mammals, and modifications probably related to the burrowing habits of modern monetremes. Their shoulder girdles are complex, including the standard components of modern mammals (scapula and clavicle), but also additional elements including coracoid, epicoracoid, and interclavicle. The scapula, however, is simplified, lacking a supraspinous fossa. The shoulder girdle is much more rigidly attached to the axillary skeleton than in other mammals. Femur and humerus are held roughly parallel to the ground when the animal walks, more in the fashion of therapsids and most modern reptiles than like modern mammals. Ribs are found on the neck (cervical) vertebrae as well as the chest (thoracic) vertebrae; in all other modern mammals, they are restricted to the thoracic region.

Another interesting skeletal characteristic of monotremes is the large epipubic bones in the pelvic region. Epipubic bones were originally thought to be related to having a pouch, but they are found in both males and females. They also occur in all species of marsupials, whether a pouch is present or not (not all marsupials have a pouch). It is now thought that epipubic bones are a vestige of the skeleton of therapsids, providing members of that group with extra attachments for abdominal muscles to support the weight of the hindquarters.

Monotremes are endothermic, but they have unusually low metabolic rates and maintain a body temperature that is lower than that of most other mammals.

The eggs layed by monotremes are small (13-15 mm diameter) and covered by a leathery shell. The number of eggs laid is small, usually 1-3, and they are placed in the mother's pouch. They contain a large yolk, which is concentrated at one end of the egg very much like the yolk of a bird's egg. Only the left ovary is functional in the platypus, but both produce eggs in the echidna. Like the eggs of birds, monotreme eggs are incubated and hatched outside the body of the mother. Incubation lasts about 12 days. The young, which are tiny and at a very early stage of development when they hatch, break out of the eggs using a "milk tooth. They are protected in a temporary pouch in echidnas but not platypuses. They are fed milk produced by mammary glands; the milk is secreted onto the skin within the pouch and sucked or lapped up by the babies. Weaning takes place when the young are 16-20 weeks old.

All male monotremes have spurs on their ankles that are presumed to be used in fighting and in defense. In one family (Ornithorhynchidae), a groove along the spur carries poison secreted by adjacent glands.

Monotremes are restricted to Australia and New Guinea. Their fossil record is very poor; the earliest fossil attributed to this group is from the early Cretaceous. A fossil from Argentina suggests that the monotremes were more widely distributed early in their history.

Evolution of Mammals

The changes from a reptilian to a mammal's lifestyle is significant. Reptiles are cold-blooded, relying on external heat sources to raise their body temperature to a point necessary for activity. During cold periods and at night, the reptile must accept substantial down time, awaiting the warmth of the next day to get started again. This leads to a lifestyle of relatively short bursts of energy needed to catch its prey and long periods of inactivity while the body processes the food. Mammals, on the other hand, are warm-blooded, generating their heat internally, retaining it by insulation of fur and/or fat. Having a more consistent body temperature means a more continuous activity level, which means more fuel (food) is required to supply the energy necessary for this active lifestyle. A more efficient metabolism is required which includes greater oxygen intake abilities and nutrient uptake. Larger lungs, diaphragm, a four-chambered heart, and improved circulatory systems were the result. A more efficient method of removing metabolic wastes was also required, resulting in improvements to the kidneys and the separation of the urinary and fecal excretory tracts (which are still combined in the monotremes, which include the platypus and echidna, the opossums and some species of shrews today).

Additionally, reptiles relied on giving birth to eggs. Eggs are basically left to hatch and fend for themselves. The new mammals evolved a system of reproduction which leaves the young to develop more fully inside the mother. After birth, the young are afforded extended parental care by the the reliable provision of the mother's mammary glands. Thus, the young can spend its energy in further physiological growth, rather than having to be immediately able to hunt and defend itself. Both these advances allowed more time for the development of a more complex organism and the cerebral system in the mammalian ancestry.

While we recognize that the "Age of Mammals" begins after the demise of the dinosaurs some 65 million years ago, the story of mammalian evolution begins well before their reign. One has to go back to a period 250 million years ago when the transition to mammals began in the form of mammal-like reptiles.

Mammals evolved from a group of reptiles called the synapsids. These reptiles arose during the Pennsylvanian Period (310 to 275 million years ago). A branch of the synapsids called the therapsids appeared by the middle of the Permian Period (275 to 225 million years ago). It was over millions of years that some of these therapsids would evolve many features that would later be associated with mammals.

It's impossible to know from which of the reptiles the first mammals evolved. The hallmarks of today's mammals - hair, warm blood and milk-producing glands - do not fossilize. We know that some dinosaurs, such as the Stegosaurus and Dimetrodon, supported large plates or sails of skin, developed to function as solar panels, absorbing the sun's heat for quick mobilization. However, during the period of dinosaurs, some of the sail-backed pelycosaurs (which includes the Dimetrodon) lost their sails. Even acknowledging a warming period ensued, it would be improbable that such an effective heating mechanism would be lost through evolution unless a more efficient method of heat control had evolved to replace it. It is therefore reasonable to supposed that the pelycosaurs and their successors, the therapsids, were to some degree endothermic. The therapsids, being only three feet long, would need some form of body insulation if it was to be effective. Thus, it is reasonable to conclude that some of these creatures were covered in hair, or fur.

It is felt that hair was first evolved in early mammals to help regulate temperatures. The earliest mammals grew these sensory cones between the scales of the evolving reptiles, which, when brushed on objects, gave a stimulus to the brain. Certain remnant scales still exist on rats' tails, armadillo shells and the backs of pangolins. Animal whiskers are still sensory receptacles. Hair is mainly the protein keratin, a form of skin, which is the same material making up the epidermis, feathers, fingernails, horns, hooves and claws. Humans have about 100,000 hairs on our head, while sea otters have 170,000 to 1,000,0000 hairs in their under fur; only one of the otters' three types of hair. Polar bear hairs have hollow cores. The cores scatter light, making the hairs seem white. Actually, they are colorless. Sunlight passes through the bear's long guard hairs to its black skin, which absorbs the radiation as heat. The inner insulating hair prevents the heat from being radiated back to the air (similar to glass of a greenhouse). Most mammals can cause their hair to become erect, thus increasing the air, (i.e., heat) retention. Although humans have lost most of their body hair, goose bumps are the results of our own relic muscles that still exist which used to control these hairs. White-tailed deer grow winter coats four times thicker than summer coats.

Early in the evolution of the therapsids arose a group called the cynodonts. These early mammal-like reptiles changed their teeth from being designed for catching and holding prey and then swallowing whole, to adding specialized teeth, including molars, designed for better mastication of food allowing for quicker digestion. In reptiles, all the teeth are alike, being replaced alternately along the jaw in waves from the back of the jaw to the front. Larger, but similar teeth replace the earlier teeth. As the cynodonts progressed toward today's mammals, this process was replace by the characteristic pattern of one set of deciduous teeth followed by one set of permanent teeth. The permanent nature of teeth enabled a specialization of teeth to serve specific functions. Additionally, the jaw of the cynodonts reduced the number of jaw bones. This freed up the superfluous bones to evolve into an entirely new function, becoming the tympanic, malleus and incus of the mammal's inner ear. Improved hearing gave these creatures a better awareness of their environment and, in turn, this increasing sensitivity called for a greater capacity for processing the auditory information in the brain.

The cynodonts dominated the world of 250 million years ago. At this time, all the land masses were joined together in one super-continent, called Pangaea, consisting of the northern Laurasia (Europe, Asia and North America) and the southern Gondwanaland (South America, Africa, Antarctica and Australia). By about 200 million years ago, the dynasty of the cynodonts suddenly was ended by the appearance of dinosaurs. Both the numbers of species and the size of the remaining species was markedly reduced. For the next 135 million years, these early mammals played a very small role in the world (both literally and figuratively), often termed the 'Dark Ages' of mammalian evolution. However, the evolution of the remaining mammals did not stop.

With the days dominated by the dinosaurs, these early mammals confined themselves to a nocturnal lifestyle or limited rocky or underground habitats. Such a lifestyle could only be possible with a warm-blooded creature, who could maintain a body temperature necessary for nocturnal activity. Being small, the physical requirements of getting around necessitated the development of agility and coordination. Also, being carnivorous added to need for augmented sensory perception. Such activities require a higher metabolism, meaning more food. Finally, being nocturnal resulted in eyes becoming relatively large, hearing more acute, the nose more sensitive, vocal cords more developed, and whiskers becoming more pronounced. All these factors combined to give the small early mammals an awareness of their environment which demanded constant fine-tuning of the integration between cerebral and physical equipment. This enabled an opportunity for increasing evolution of reproductive strategies, maternal behavior, parental care, communication between individuals and learning.

In a manner of speaking, the net result was that, while dinosaurs were getting larger and developing extensive defensive and offensive tools, they were really evolving primarily their physical hardware, while the evolution of the early mammals used its energy in evolving their brain and behavioral software.

During this period of dinosaur dominance, the three forms of mammals that still exist today appeared; the monotremes, the marsupials, and the placentals. All three of these groups are believed to have originated from a common ancestor. If one were to define the first mammal as the most recent common ancestor of these three forms of mammals, as some do, then an animal found in England (Phascolotherium bucklandi) lays claim to that honor, roaming some 165 million years ago during the middle Jurassic period. Others argue for a creature from Texas during the late Triassic, about 225 million years ago, with the name of Adelobasileus cromptoni. By the Jurassic Period (180 to 130 million years ago), mammals were shrew-like animals, being insectivorous carnivores.

The earliest mammals are generally agreed to have been the ancestors of todays' monotremes. Monotremes resemble reptiles and differ from all other mammals in that they lay shell-covered eggs that are incubated and hatched outside of the body of the mother. Like reptiles, the ducts of the excretory system and the genital ducts open into a single external opening known as the cloaca (thus, the ordinal name Monotremata). Like mammals, they have fur, a four-chambered heart, are warm-blooded, and nurse their young from specialized glands. However, as might be expected in these primitive mammals, the four-chambered heart is, to an extent, incomplete, the body temperature averages lower than that of other mammals, and their are no nipples; rather, the milk is excreted through skin glands, with the young suckling the body fur to receive the milk. The monotremes are represented today by two species of echidnas and one species of platypus.

These monotremes split from the early mammals around 130 million years ago, probably in Australia, while it was still joined to Antarctica, as part of Gondwanaland. Monotremes had dispersed through Antarctica into South America by the early Paleocene (65 million years ago), but are only found today in Australia, where they were safe from competition with the more advanced placental mammals.

About 120 million years ago, the mammalian line ceased laying eggs and began bearing live young. These forms of mammals were the first marsupials, who bore their young at a very early stage in their development and transferred them to a pouch where modified sweat glands secreted milk. It is generally accepted that the first marsupials arose in North America and spread to South America, then to Antarctica and Australia some time before the breakup of Pangaea near the end of the Cretaceous period, some 70 million years ago. Others argue that a southern continental origin is more probable.

By the end of the Cretaceous (65 million years ago), Pangaea had begun to break apart. These early marsupials (and the limited number of monotremes) were the sole inhabitants of the southern landmass, Gondwanaland, and the marsupial lines radiated out to fill out every niche available. (Although later continental movements enabled carnivorous placental mammals to reach South America and Africa, the islands of Antarctica and Australia remained to be occupied solely by marsupials and monotremes.

Meanwhile, in Laurasia, another branch of mammals took the marsupial reproductive strategy one step further. As the early mammals evolved, they continued to expand their awareness of the environment reflected in improved senses of sight, smell, touch and hearing. These processes, and an increasing need for coordination, required constantly more cerebral capacity. The placental evolution best suited this need. In this system, the embryo remained in the uterus, receiving nutrients and oxygen from the mother for an extended amount of time, enabling the development of the brain. As the brain continued to expand, it brought the mammals greater ability to learn, continuing the cycle and speeding up the process well beyond the levels attained by marsupials. The splitting of the placentals from the marsupials date back to about 120 million years ago. The first living placental order is believed by most scientists to be the Insectivores (including shrews and moles), while others support the Xenarthra, which includes the armadillos, sloths, and anteaters as more primitive.

About 70 million years ago, placentals and marsupials arrived in South America; the placentals being the herbivores, and the marsupial, the carnivores.

With the lowering ocean level creating the Panama land bridge about 3 million years ago, the placental carnivores (including the saber-toothed tigers) migrated south from North America and wiped out the marsupial carnivores. This land bridge was responsible for the northward migration of opossum, armadillos and porcupines and the southward migration of raccoons, weasel, dogs, bear and cats (including the saber-toothed cat). Mastodons, tapirs, horses, llamas and deer were the hoofed emigrants southward. Rabbits and rodents (including the very successful mice) also went south. Today, fully half of the land mammal genera of South America came by way of this land bridge, while only three genera (the porcupine, opossum and armadillo) having found permanent homes in North America from the south.

By the end of the Cretaceous period, some 65 million years ago, as the continents diverged, the climate became more erratic, the vegetation changed, and the dinosaurs disappeared, the 'Age of Mammals' finally gets under way.

So many changes were occurring. Of course, the removal of the dinosaurs opened the daylight world to the mammals. Flowering plants, which had quickly dominated the landscape over the coniferous gymnosperms, evolved symbiotic relationships with the mammals, bringing mutual benefit to both plants and animals; food for the animals and a method of fertilization for the plants, thus enabling a radiation of new species of both plants and animals. And the movement of the continental plates caused the formation of new mountains and new habitats. As the Atlantic basin widened, the Andes and the Rocky Mountains rose where the American continents independently butted up against the Pacific plate. India broke free from Africa and drifted north towards Asia, creating the Himalayan range. As Africa turned a little and pushed against Europe, both continental plates crumpled a little, creating the Alps and the Atlas Mountains. With these new habitats came new plants, insects, and soon, new mammals.

The first mammalian carnivores to evolve from (and to feed on) the ancestral insectivores are known as the creodonts, which split into two main groups about 55 million years ago; the felids and the canids, or as we know them, the cats and dogs.

Next came the herbivores to feed on the plant material itself. The teeth became broad, extremely hard and high-crowned to grind the tough fibrous plant material. The digestive system became specially adapted for processing low-nutrient plants, and the unnecessary claws evolved into hooves, which in Latin is ungula, thus, the name of ungulates. Like the carnivores, these herbivores split into two main groups; the odd-toed perissodactyls and the even-toed artiodactyls.

As has been the case throughout the evolution of animals, where food resources will allow, gigantic forms arose, and where food was limited smaller forms predominated. The concept of convergent, or parallel, evolution led each continent to evolve similar types of cats, dogs, rodents, pigs, and rhinoceros to fill in similar ecological niches.

As temperatures moderated from the Paleocene into the Eocene (from 53 to 36 million years ago), tropical forests aided in the rise and spread of the flowering plants and the radiation of the mammals. But, throughout the Oligocene and into the Miocene (from 36 to 8 million years ago), the temperature dropped about 5 °F as a result of a shift in the axis of the Earth's rotation and resulting formation of the ice cap on Antarctica, which at this time was arriving at the south pole. This drop in temperature (and consequential drop in precipitation) caused a change in vegetation from forest to grasslands. With this gradual change in the environment, came an evolutionary adaptation among the animals. Grass replaced forest in many regions and with it a number of habitats for browsers, who primarily feed on woody plants, were lost.

However, certain animals were able to capitalize on the dense stands of grass vegetation which enabled larger animal populations than afforded by forest habitats. With the exploitation of grasses, teeth became a limiting factor in species diversity since the grasses contained crystal of silica, which quickly wore down teeth. As a result, new species with specially adapted teeth incorporating hard enamel evolved, feeding primarily on these grasses. These new animals are known as grazers.

The grass itself benefits from the presence of the herds for they trample and eat the seedling of bushes or trees that might take root on the plain and that would, were they to grow tall, rob the grass of light and eventually displace it. It seems likely therefore that the spread of the grasslands and the evolution of grazing animals proceeded together, step by step.

Grass is high in cellulose, a material hard to digest. One adaptation was the emergence of bacteria in the gut to help break down the grass. The lagomorphs, including rabbits and hares, furthered this adaptation by passing the grass through their digestive system twice. After excreting the food product, it is ingested again to extract as much of the nutrients as possible. This process is called coprophagy, and is also practiced by some mice and beaver.

Certain of the even-toed ungulates evolved an extra stomach (the rumen) which partially broke down the food before passing it back to the mouth for further chewing before being sent down to the true stomach for final digestion. This, of course, is the proverbial "chewing of the cud." These are known as the ruminants. The success of these more advanced ungulates can be shown by the number of existing species. There are 194 species of even-toed ungulates and only 16 odd-toed ungulates, of which the horse is the only exclusive grazer among them.

This cud chewing makes for longer time for food to be processed (70 - 100 hours through the gut) than, say, for a horse (30 -45 hours). The central difference, then, is that the digestive system of the horse is less efficient than that of the ruminant, but in compensation the horse eats greater quantities of food; the emphasis in ruminants is on highly efficient digestion and on more selective feeding, but not on high rates of food intake. Given food in short supply, the ruminant will probably survive after the horse has starved to death.

Within the ruminants, antlers and horns are found in the most progressive families. The antlered artiodactys include the deer, elk, caribou and moose. Antlers are usually present and occur in males only, with the exception of caribou, where both sexes have antlers. Antlers differ from horns in being annually shed and regrown each year. In the Giraffadae (giraffe and okapi), both male and female have fur covered bone protuberances. In the Antilocapridae (pronghorn sheep), both male and female have a boney spur with an annually shed sheath. Finally, in the Bovidae family (antelope, bison, sheep, goats, and cattle), usually both male and female have horns. The horns are never shed and in some species grow throughout the life of the animal. Bovid horns are never branched, but can be spectacularly curved or spiraled. The domestication of bovids about 8,000 years ago was one of the major factors responsible for the first civilization in the Tigris/Euphrates River valley in Mesopotamia (annual grain crops, including wheat and barley were another reason). A few bovids reached the New World in the Pleistocene via the Bering Strait land bridge. Because this avenue of dispersal was under the influence of severe boreal climatic conditions at this time, it functioned as a filter bridge, and only animals adapted to these conditions dispersed across the bridge. Migrators to the New World included bighorn sheep, mountain goat, musk ox, and the bison. Other Old World bovids that couldn't make the cold crossing included antelopes and the gazelles. Bovids inhabit grasslands, making their greatest stand currently in the savannas and grasslands of Africa.

In addition to the ungulates, the rodents prospered as plant eaters. They, too, dealt with the tough plant material. Their evolution was to maintain open roots to their front gnawing teeth, the incisors, so that they continue to grow throughout the animal's life, compensating for wear. Like the ungulates, rodents also enlisted the help of bacterial life to further break down the cellulose in the guts of the animals.

And, of course, with the grazers coming out in the open grasslands, came the predators. Only the largest of the herbivores (elephants and rhinoceros) were safe from predation. The smaller rodents burrowed for protection. Others chose speed to outrun predators. Horses, for example, evolved in North America, running on its' toes, which, over time, were reduced from four toes to one. (Numerous toes, necessary for better traction in the soft soil of the forests, became unnecessary, and, in fact, a liability on the hard ground of the drier grasslands.) Legs increased in length for running and the skull elongated to enable the eyes to better see predators while they fed.

As part of the mixing of continental members 3 million years ago, horses traveled over the Bering Strait to the Old World and over the Panama isthmus to South America. Dramatic climatic shifts and the arrival of man contributed to the extinction of Equus in the Americas. Only one genus world-wide exists in the Old World from their maximum in the Miocene. This genus, includes all the world's horses; zebras, asses and onagers - the ancestor of the domesticated horse. They spread across the Bering Strait into Europe and then Africa, evolving into today's horses, donkeys and zebras, before dying out in North America.

Speed was the name of the game. While the ungulates evolved speed by running on their toes, the predators couldn't do this since they needed their claws for the catch. Instead, they effectively lengthened their limbs by making their spine extremely flexible, enabling their hind and front legs to overlap beneath the body. Or they hunted in groups, like the lions and dogs.

Back in the forests, some of the tiny insectivorous creatures from the days in the shadows of the dinosaurs, adapted a lifestyle in the trees. The foods available in the trees were rich and varied, including leaves, buds, fruit, insects, eggs and nestlings. As opposed to a terrestrial life, where the sense of smell was the most important sensory organ, life in the trees emphasized the development of vision. Eyesight evolved to include looking sideways, integrating the two fields of vision to give a stereoscopic image of the surrounding, with perhaps even the first tints of color. In addition, agility and coordination were critical. Use of the forehands to grasp branches and food resulted in opposable thumbs, and the rear legs became more the means of propulsion and static support, leading ultimately to the ability to sit with the back erect. In addition to these traits, an omnivorous diet required a range of teeth functions, not just the slicers of the carnivores nor the grinders of the grazers. Along with these developments, was the co-committal development of the cerebral capacity. From this beginning, arose the primates.

With the origin of the primates, some 80 million years ago, came certain features that would forever differentiate these mammals from all others. Among them, they have retained four kinds of teeth (incisors, canines, premolars and molars), opposable thumbs, two mammary glands, frontally directed eyes and a relatively large brain. It is from this stock that has lead to the evolution of the human being, another story in itself.

Mammalian Reproduction

You probably realize that cats, dogs, people, and other mammals don't typically lay eggs. There are exceptions, however. Egg-laying is possible among the monotremes, mammals with bird-like and reptilian characteristics. Recall that mammals can be classified into three general groups, based on their reproductive strategy: the monotremes, the marsupials and the placental mammals.

The egg-laying monotremes, such as echidnas (figure below) and platypuses (figure below), use one opening, the cloaca, to urinate, release waste, and reproduce, just like birds. They lay leathery eggs, similar to those of lizards, turtles, and crocodilians. Monotremes feed their young by "sweating" milk from patches on their bellies, as they lack the nipples present on other mammals.

The echidna (right) is a member of the monotremes, the most primitive order of mammals. Another monotreme, the platypus (left), like other mammals in this order, lays eggs and has a single opening for the urinary, genital, and digestive organs.

All other mammals give birth to live young and belong to one of two different categories, the marsupials and the placental mammals. A marsupial is an animal in which the embryo, which is often called a joey, is born at an immature stage. Development must be completed outside the mother's body.

Most female marsupials have an abdominal pouch or skin fold where there are mammary glands. The pouch is a place for completing the development of the baby. Although blind, without fur, and with only partially formed hind legs, the tiny newborns have well developed forelimbs with claws that enable them to climb their way into their mother's pouch where they drink their mother's milk and continue their development. Marsupials include kangaroos, koalas, and opossums. Other marsupials are the wallaby and the Tasmanian Devil. Most marsupials live in Australia and nearby areas.

The majority of mammals are placental mammals. These are mammals in which the developing baby is fed through the mother's placenta. Female placental mammals develop a placenta after fertilization. A placenta is a spongy structure that passes oxygen, nutrients, and other useful substances from the mother to the fetus. It also passes carbon dioxide and other wastes from the fetus to the mother. The placenta allows the fetus to grow for a long time within the mother.

A marsupial mammal, this eastern gray kangaroo has a joey (young kangaroo) in its abdominal pouch.

Some mammals are alone until a female can become pregnant. Others form social groups with big differences between sexes, such as size differences, a trait called sexual dimorphism. Dominant males are those that are the largest or best-armed. These males usually have an advantage in mating. They may also keep other males from mating with females within a group. This is seen in elephant seals (Figure below), and also with elk, lions and non-human primates, including the orangutans and gorillas. Male elk grow antlers, while female elk do not have antlers. Adult male lions are not only larger than females, they have a mane of long hair on the side of the face and top of the head.

A mating system with a group of many females and one male, as seen in the seal species. Male elephant seals can grow to 14 feet long, whereas females can grow to 11 feet long.

Reproductive System

Placental Mammals

- Male placental mammals

The mammalian male reproductive system contains two main divisions, the penis and the testicles, the latter of which is where sperm are produced. In humans, both of these organs are outside the abdominal cavity, but they can be primarily housed within the abdomen in other animals. For instance, a dog's penis is covered by a penile sheath except when mating. Having the testicles outside the abdomen best facilitates temperature regulation of the sperm, which require specific temperatures to survive. The external location may also cause a reduction in the heat-induced contribution to the spontaneous mutation rate in male germinal tissue. Sperm are the smaller of the two gametes and are generally very short-lived, requiring males to produce them continuously from the time of sexual maturity until death. The produced sperm are stored in the epididymis until ejaculation. The sperm cells are motile and they swim using tail-like flagella to propel themselves towards the ovum. The sperm follows temperature gradients (thermotaxis) and chemical gradients (chemotaxis) to locate the ovum.

- Female placentals

The mammalian female reproductive system likewise contains two main divisions: the vagina and uterus, which act as the receptacle for the sperm, and the ovaries, which produce the female's ova. All of these parts are always internal. The vagina is attached to the uterus through the cervix, while the uterus is attached to the ovaries via the Fallopian tubes. At certain intervals, the ovaries release an ovum, which passes through the fallopian tube into the uterus.

If, in this transit, it meets with sperm, the egg selects sperm with which to merge; this is termed fertilization. The fertilization usually occurs in the oviducts, but can happen in the uterus itself. The zygote then implants itself in the wall of the uterus, where it begins the processes of embryogenesis and morphogenesis. When developed enough to survive outside the womb, the cervix dilates and contractions of the uterus propel the fetus through the birth canal, which is the vagina.

The ova, which are the female sex cells, are much larger than the sperm and are normally formed within the ovaries of the fetus before its birth. They are mostly fixed in location within the ovary until their transit to the uterus, and contain nutrients for the later zygote and embryo. Over a regular interval, in response to hormonal signals, a process of oogenesis matures one ovum which is released and sent down the Fallopian tube. If not fertilized, this egg is released through menstruation in humans and other great apes, and reabsorbed in other mammals in the estrus cycle.

- Gestation

Gestation, called pregnancy in humans, is the period of time during which the fetus develops, dividing via mitosis inside the female. During this time, the fetus receives all of its nutrition and oxygenated blood from the female, filtered through the placenta, which is attached to the fetus' abdomen via an umbilical cord. This drain of nutrients can be quite taxing on the female, who is required to ingest slightly higher levels of calories. In addition, certain vitamins and other nutrients are required in greater quantities than normal, often creating abnormal eating habits. The length of gestation, called the gestation period, varies greatly from species to species; it is 40 weeks in humans, 56–60 in giraffes and 16 days in hamsters.

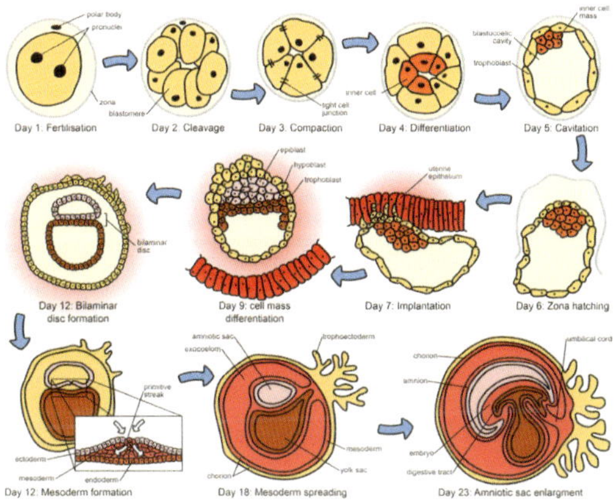

The initial stages of human embryogenesis.

- Birth

Once the fetus is sufficiently developing, chemical signals start the process of birth, which begins with contractions of the uterus and the dilation of the cervix. The fetus then descends to the cervix, where it is pushed out into the vagina, and eventually out of the female. The newborn, which is called an infant in humans, should typically begin respiration on its own shortly after birth. Not long after, the placenta is passed as well. Most mammals eat this, as it is a good source of protein and other vital nutrients needed for caring for the young.

Monotremes

Monotremes, only five species of which exist, all from Australia and New Guinea, are mammals that lay eggs. They have one opening for excretion and reproduction called the cloaca. They hold the eggs internally for several weeks, providing nutrients, and then lay them and cover them like birds. Like marsupial "joeys", monotreme "puggles" are larval and fetus-like, as like them they cannot expand their torso due to the presence of epipubic bones, forcing them to produce undeveloped young.

Marsupials

Marsupials' reproductive systems differ markedly from those of placental mammals, though it is probably the plesiomorphic condition found in viviparous mammals, including non-placental eutherians. During embryonic development, a choriovitelline placenta forms in all marsupials. In bandicoots, an additional chorioallantoic placenta forms, although it lacks the chorionic villi found in eutherian placentas.

Gametogenesis

Animals, including mammals, produce gametes (sperm and egg) through meiosis in gonads (testicles in males and ovaries in females). Sperm are produced by the process of spermatogenesis and eggs are produced by oogenesis. These processes are outlined in the article gametogenesis. During gametogenesis in mammals many genes encoding proteins that take part in DNA repair mechanisms show enhanced or specialized expression. These mechanisms include meiotic homologous recombinational repair and mismatch repair.

Pregnancy (Mammals)

Mammalian pregnancy encompasses everything that occurs between fertilization and birth and involves among others internal fertilization, the retention of the developing egg within the female reproductive tract and direct maternal nourishment (matrotrophy).

The first mammals were egg-laying, and, evolutionarily speaking, pregnancy is a relative newcomer. Pregnancy has been described as the superimposition of live birth (viviparity) on mammalian biology. The repurposing of the fetal membranes of egg-laying vertebrates into the placenta of viviparous mammals or the degeneration and loss of yolk genes are just two of the many hallmarks of this remarkable evolutionary transition.

Advantages of Pregnancy

Numerous benefits of pregnancy have been proposed, from protection against thermal extremes, osmotic stress or predation, to the advantages associated with unremitting provisioning of offspring. But there are clear costs associated with pregnancy as well, including increased maternal energy expenditure, reduction in female mobility and fecundity, and risk of injury to both mother and fetus from immunological and inflammatory responses inutero, or inadvertent exchange of molecules and cells. It is unlikely that any single factor accounts for the evolution of pregnancy.

Not all Mammals get Pregnant

The monotreme mammals (think of the platypus) are egg-laying. Monotremes nourish their embryos with yolk, but they also have a transient embryonic structure for maternal and fetal exchange of nutrients and gases, which some consider a rudimentary placenta. In contrast, marsupials (kangaroos and kin) and eutherians (all the rest, including us) give birth to live young. Marsupial and eutherian embryos are also nourished by placentas, but these are much more complex in structure and function; however, the two lineages differ in the particular mix of membranes they use to form their placentas. Of the many important differences between mammals, the length of gestation is a big one: gestation in marsupials and monotremes is extremely short relative to adult lifespan, and much of development is completed after birth with the help of mother's milk. Eutherians lactate as well, but gestation is much longer and their offspring are much further developed when born.

Variation within Eutherian Mammals

Eutherians express a bewildering range of gestational traits, particularly so in the placenta, arguably the most structurally variable and functionally diverse eutherian organ. For example, humans have a very invasive, disk-shaped placenta that is bathed directly in maternal blood. And your canine or feline friend, with whom you may share a house or perhaps a bed? A completely different placenta in almost every respect. It is as if you were interested in planting a garden, and as you shopped for plants, you discovered that nearly every type of plant required water from its own peculiar garden hose.

Reasons of Variations

Placental diversity is organized along several interrelated dimensions having to do with its structure and the relationship between maternal and fetal circulatory systems. One long-standing idea has been that variation in the maternofetal interface must reflect the differing physiological needs of pregnancy across organisms. Certain placental types might be better at extracting maternal resources, for example, and groups with energetically 'expensive' pregnancies might require such placentas. But it turns out that it is not quite so simple. Recent phylogenetic analyses indicate that, in some eutherian mammals, what might seem to be more efficient placentae for nutrient transfer have been replaced with placentae that are wellseparated from the maternal circulation, or have a smaller area of contact with maternal tissues.

Pregnancy Optimized for Reproduction

It is only during pregnancy that the cells of two genetically distinct individuals come into such prolonged and intimate proximity. Live birth opens the door for 'negotiations' between the developing fetus and the mother over provisioning in utero, something not possible in oviparous species. Since viviparity requires a significant investment of maternal resources into fetal growth, conflict can occur because the fetus only shares half of its genes with its mother. In certain situations, the paternally and maternally derived genes in the fetus can 'disagree' with the amount of maternal provisions the fetus receives, favoring greater investment than the mother would otherwise provide.

Evidence for Conflicts in Pregnancy

An unusual fraction of genes expressed in the eutherian placenta is imprinted, and in particular genes involved in nutrient transfer. Imprinted genes show bias in the expression of one allele over the other based on the parent each allele came from. For these imprinted genes, alleles inherited from the dad tend to promote fetal growth, whereas those from the mom tend to control it. Genomic imprinting apparently is not present in monotremes, and seems to have evolved in concert with placentation in marsupials and eutherians. In this light, the evolution of noninvasive placentation in organisms with expensive investment strategies might be seen as a means to manage the perils of fetal demands. Less invasive placentation transfers greater control over investment to the pregnant mother; in effect, a small step back towards the relative conflict-free bliss of egg-laying.

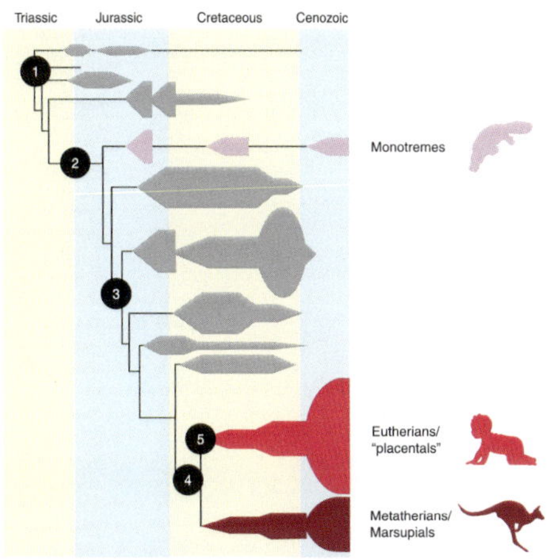

Cartoon depiction of the likely origins of key aspects of mammalian pregnancy.

1. Triassic origin of mammals; lactation.
2. Origin of matrotrophy in egg-laying mammals.
3. Origin of viviparity; origin of maternal fetal conflict.
4. Origin of mammalian placenta; origin of genomic imprinting.
5. Origin of eutherian placenta; much longer gestation.

Pregnancy clearly works most of the time, and there are ingenious feats of cooperation at the interface between mother and fetus. For example, following implantation, the uterine lining of some mammals forms a temporary tissue barrier on which the fetal placenta will attach and grow. This barrier nourishes the embryo, regulates the interaction between maternal and fetal tissues, while protecting the fetus from the maternal immune system. In some mammals, such as elephant shrews, bats, and the great apes (including us), this process occurs spontaneously, in anticipation of implantation; when fertilization does not take place, the consequence is menstruation. On the fetal side, specialized trophoblast cells with secretory properties fuse to form an unbroken multinucleate outer sheath or syncytium that entirely envelops the placenta, helping to protect it and the developing fetus from the maternal immune system. Without these cells, both mother and fetus would be at risk of a damaging immune reaction. Amazingly, formation of this sheath comes courtesy of domesticated viral genes that express cell fusion-promoting proteins in fetal trophoblast cells, helping to convert these into a syncytium.

Human babies have big brains when they are born. However, relative to adult brain size, their brains are actually smaller than other primates, and consequently, at the time of birth they are behaviorally at an earlier developmental stage than what might seem optimal. Shouldn't human pregnancy be longer than nine months? Evolutionary anthropologists have long hypothesized that this 'obstetric dilemma' has something to do with our bipedalism and how it constrains the pelvic birth canal. The relative immaturity of human babies, they hypothesize, is a solution to the trade-off between having a big brain and walking upright. A competing theory is that birth timing is set by the point at which it becomes metabolically inefficient or even impossible for the mother to nurture the fetus within her uterus, relative to the provisioning she could accomplish after birth. Whatever the explanation, the factors that determine birth timing are clearly important in humans; pregnancy complications such as pre-eclampsia, pre-term birth, and spontaneous abortion have all been associated with defective placentation and dysregulation of birth timing. A key to understanding and solving some of the most pressing health issues in human pregnancy lies in a better understanding of how pregnancy evolved, and the adaptations, constraints, and conflicts that have shaped mammalian gestation.

References

- What-is-mammal, animals-and-nature: dkfindout.com, Retrieved 31 March 2018
- Classification-of-mammals: animalsworlds.com, Retrieved 30 June 2018
- Mammalia, Metatheria-characters-and-distribution-chordata-zoology-8893: notesonzoology.com, Retrieved 11 July 2018
- Mammalogy, science: britannica.com, Retrieved 23 May 2018
- Evolution-of-mammals: bobpickett.org, Retrieved 18 July 2018
- Mammal-Reproduction-MS-LS, mammal-reproduction, biology: www.ck12.org, Retrieved 28 June 2018

Chapter 2

Large Carnivorous Mammals

Some large carnivorous mammals are tiger, lion, leopard, jaguar, cheetah, puma, bear, etc. The topics elaborated in this chapter provide a detailed description of such large carnivores for a holistic understanding of the subject.

Tiger

Tigers belong to the First largest of the cat species.

Size

Tigers have distinctive stripes, which help camouflage them when hunting prey. Some tigers have orange fur with black stripes; others are black with tan stripes, white with tan stripes or all white (albino). No two tigers have the same markings on their coats. They are as individual as fingerprints are for humans.

On average, tigers are 4.8 to 9.5 feet (1.5 to 2.9 m) long and weigh 165 to 716 lbs. (75 to 325 kilograms). The largest tigers, the Siberian, also called Amur, are 10.75 feet (3.3 meters) long and weigh 660 lbs. (300 kg), according to National Geographic. The smallest tiger is the Sumatran tiger. They grow to 5 to 12 feet (1.5 to 3.7 m) and weigh 143-670 lbs. (65-305 kg), according to the Woodland Park Zoo. Tigers also have very long tails, which can add 2.3 to 3.6 feet (0.7 to 1.1 m) to their overall length.

Habitat

Tigers live in Asia. Larger subspecies, such as the Siberian tiger, tend to live in northern, colder areas, such as eastern Russia and northeastern China. Smaller subspecies live in southern, warmer countries, such as India, Bangladesh, Nepal, Bhutan, Myanmar, Laos, Cambodia, Vietnam, Malaysia and Indonesia.

They live in arid forests, flooded mangrove forests, tropical forests and taiga, depending on the subspecies.

Diet

All tigers are carnivores. Most of a tiger's diet consists of large prey, such as pigs, deer, rhinos or elephant calves. To kill their prey, tigers will clamp down on the animal's neck with their jaws and suffocate the animal. Though tigers are fierce hunters, they are no strangers to failure. Ninety percent of the time they don't catch their prey, according to the Woodland Park Zoo.

Habits

Tigers are solitary creatures; they like to spend most of their time alone, roaming their massive territories looking for food. According to the San Diego Zoo, the Siberian tiger has the largest range. Its territory can be more than 4,000 square miles (10,000 square kilometers). Tigers mark their territory by scratching marks into trees with their claws.

Offspring

Tiger babies, or cubs, are born helpless. At birth, a cub weighs 2.2 pounds (1 kg), and a female may have as many as seven cubs at a time. Usually, only two survive, though, because the mother must leave the cubs while she hunts, and she cannot kill enough prey to feed so many cubs.

Tiger cubs mature quickly. At 8 weeks old, they are ready to learn how to hunt and go out on hunting expeditions with their mother. At 2 years old, the cubs will set out on their own, and their mother will have another set of cubs. Tigers typically live 14 to 18 years.

Conservation Status

There aren't a lot of tigers left in general, though. With only six subspecies left, tigers have been declared endangered by the IUCN. Tigers that remained in the forests are only a little more than 3,000 in number. The rest of the tigers, which may have not been included in the count, belong to small, isolated groups that sort of act like family units. The population has suffered a rapid decline due to the destruction of the tigers' homes as well as because of the crime of poaching. So, at this point, the area left inhabited by tigers is at only 184,911 square kilometers or 457,497 square miles. This estimate, however, was good in the 1990s. Today, the area could be a lot smaller.

Conservation

Major threats to the tiger include habitat destruction, habitat fragmentation and poaching for fur and body parts, which have simultaneously greatly reduced tiger populations in the wild In India, only 11% of the historical tiger habitat remains due to habitat fragmentation. Demand for tiger parts for use in traditional Chinese medicine has also been cited as a major threat to tiger populations. At the start of the 20th century, it was estimated there were over 100,000 tigers in the wild, but the population has dwindled outside of captivity to between 1,500 and 3,500. Some estimates suggest that there are fewer than 2,500 mature breeding individuals, with no subpopulation containing more than 250 mature breeding individuals. The global wild tiger population was estimated by the World Wide Fund for Nature at 3,200 in 2011 and 3,890 in 2015—*Vox* reported that this was the first increase in a century.

Tiger population status (2016)	
Country	**Estimate**
Bangladesh	106
Bhutan	103
Cambodia	0
China	>7
India	2,226
Indonesia	371
Laos	2
Malaysia	250
Myanmar	no data
Nepal	198
Russia	433
Thailand	189
Vietnam	<5
Total	**3,890**

India is home to the world's largest population of wild tigers. A 2014 census estimated a population of 2,226, a 30% increase since 2011. In 1973, India's *Project Tiger*, started by Indira Gandhi, established over 25 tiger reserves in reclaimed land, where human development was forbidden. The project was credited with tripling the number of wild Bengal tigers from some 1,200 in 1973 to over 3,500 in the 1990s, but a 2007 census showed that numbers had dropped back to about 1,400 tigers because of poaching. Following the report, the Indian government pledged $153 million to the initiative, set up measures to combat poaching, promised funds to relocate up to 200,000 villagers in order to reduce human-tiger interactions, and set up eight new tiger reserves. India also reintroduced tigers to the Sariska Tiger Reserve and by 2009 it was claimed that poaching had been effectively countered at Ranthambore National Park.

In the 1940s, the Siberian tiger was on the brink of extinction with only about 40 animals remaining in the wild in Russia. As a result, anti-poaching controls were put in place by the Soviet Union and a network of protected zones (zapovedniks) were instituted, leading to a rise in the population to several hundred. Poaching again became a problem in the 1990s, when the economy of Russia collapsed. The major obstacle in preserving the species is the enormous territory individual tigers require (up to 450 km^2 needed by a single female and more for a single male). Current conservation efforts are led by local governments and NGO's in concert with international organizations, such as the World Wide Fund for Nature and the Wildlife Conservation Society. The competitive exclusion of wolves by tigers has been used by Russian conservationists to convince hunters to tolerate the big cats. Tigers have less impact on ungulate populations than do wolves, and are effective in controlling the latter's numbers. In 2005, there were thought to be about 360 animals in Russia, though these exhibited little genetic diversity. However, in a decade later, the Siberian tiger census was estimated from 480 to 540 individuals.

Wild Sumatran Tiger

Having earlier rejected the Western-led environmentalist movement, China changed its stance in the 1980s and became a party to the CITES treaty. By 1993 it had banned the trade in tiger parts, and this diminished the use of tiger bones in traditional Chinese medicine. After this, the Tibetan people's trade in tiger skins became a relatively more important threat to tigers. The pelts were used in clothing, tiger-skin *chuba* being worn by singers and participants in horse racing festivals, and had become status symbols. In 2004, international conservation organizations launched successful environmental propaganda campaigns in China against the Tibetan tiger skin trade. There was outrage in India, where many Tibetans live, and the 14th Dalai Lama was persuaded to take up the issue. Since then there has been a change of attitude, with some Tibetans publicly burning their chubas.

In 1994, the Indonesian Sumatran Tiger Conservation Strategy addressed the potential crisis that tigers faced in Sumatra. The Sumatran Tiger Project (STP) was initiated in June 1995 in and around the Way Kambas National Park in order to ensure the long-term viability of wild Sumatran tigers and to accumulate data on tiger life-history characteristics vital for the management of wild populations. By August 1999, the teams of the STP had evaluated 52 sites of potential tiger habitat in Lampung Province, of which only 15 these were intact enough to contain tigers. In the framework of the STP a community-based conservation programme was initiated to document the tiger-human dimension in the park in order to enable conservation authorities to resolve tiger-human conflicts based on a comprehensive database rather than anecdotes and opinions.

The Wildlife Conservation Society and Panthera Corporation formed the collaboration *Tigers Forever*, with field sites including the world's largest tiger reserve, the 21,756 km^2 (8,400 sq mi) Hukaung Valley in Myanmar. Other reserves were in the Western Ghats in India, Thailand, Laos, Cambodia, the Russian Far East covering in total about 260,000 km^2 (100,000 sq mi).

Tigers have been studied in the wild using a variety of techniques. Tiger population have been estimated using plaster casts of their pugmarks, although this method was criticized as being inaccurate. More recent techniques include the use of camera traps and studies of DNA from tiger scat, while radio-collaring has been used to track tigers in the wild. Tiger spray has been found to be just as good, or better, as a source of DNA than scat.

The exact number of wild tigers is unknown, as many estimates are outdated or educated guesses; few estimates are based on reliable scientific censuses. The table shows estimates according to IUCN Red List accounts and range country governments dating from 2009 to April 2016.

Rewilding and Reintroduction Projects

In 1978, the Indian conservationist Billy Arjan Singh attempted to rewild a tiger in Dudhwa National Park; this was the captive-bred tigress Tara. Soon after the release, numerous people were killed and eaten by a tigress that was subsequently shot. Government officials claimed it was Tara, though Singh disputed this. Further controversy broke out with the discovery that Tara was partly Siberian tiger.

A rewilded South China tiger hunting blesbok in South Africa

The organisation Save China's Tigers has attempted to rewild the South China tigers, with a breeding and training programme in a South African reserve known as Laohu Valley Reserve (LVR) and eventually reintroduce them to the wild of China.

A future rewilding project was proposed for Siberian tigers set to be reintroduced to northern Russia's Pleistocene park. The Siberian tigers sent to Iran for a captive breeding project in Tehran are set to be rewilded and reintroduced to the Miankaleh peninsula, to replace the now extinct Caspian tigers.

Relation with Humans

Tiger Hunting

The tiger has been one of the big five game animals of Asia. Tiger hunting took place on a large scale in the early 19th and 20th centuries, being a recognised and admired sport by the British in colonial India as well as the maharajas and aristocratic class of the erstwhile princely states of pre-independence India. A single maharaja or English hunter could claim to kill over a hundred tigers in their hunting career. Tiger hunting was done by some hunters on foot; others sat up on *machans* with a goat or buffalo tied out as bait; yet others on elephant-back.

Figure(a): Tiger hunting on elephant-back, India

Figure(b): A hunting party poses with a killed Javan tiger

Historically, tigers have been hunted at a large scale so their famous striped skins could be collected. The trade in tiger skins peaked in the 1960s, just before international conservation efforts took effect. By 1977, a tiger skin in an English market was considered to be worth US$4,250.

Traditional Medicine

Many people in China and other parts of Asia have a belief that various tiger parts have medicinal properties, including as pain killers and aphrodisiacs. There is no scientific evidence to support these beliefs. The use of tiger parts in pharmaceutical drugs in China is already banned, and the government has made some offences in connection with tiger poaching punishable by death. Furthermore, all trade in tiger parts is illegal under the Convention on International Trade in Endangered Species of Wild Fauna and Flora and a domestic trade ban has been in place in China since 1993.

However, the trading of tiger parts in Asia has become a major black market industry and governmental and conservation attempts to stop it have been ineffective to date. Almost all black marketers engaged in the trade are based in China and have either been shipped and sold within in their own country or into Taiwan, South Korea or Japan. The Chinese subspecies was almost completely decimated by killing for commerce due to both the parts and skin trades in the 1950s through the 1970s. Contributing to the illegal trade, there are a number of tiger farms in the country specialising in breeding the cats for profit. It is estimated that between 5,000 and 10,000 captive-bred, semi-tame animals live in these farms today. However, many tigers for traditional medicine black market are wild ones shot or snared by poachers and may be caught anywhere in the tiger's remaining range (from Siberia to India to the Malay Peninsula to Sumatra). In the Asian black market, a tiger penis can be worth the equivalent of around $300 U.S. dollars. In the years of 1990 through 1992, 27 million products with tiger derivatives were found. In July 2014 at an international convention on endangered species in Geneva, Switzerland, a Chinese representative admitted for the first time his government was aware trading in tiger skins was occurring in China.

Man-eating Tigers

Wild tigers that have had no prior contact with humans actively avoid interactions with humans. However, tigers cause more human deaths through direct attack than any other wild mammal. Attacks are occasionally provoked, as tigers lash out after being injured while they themselves are hunted. Attacks can be provoked accidentally, as when a human surprises a tiger or inadvertently comes between a mother and her young, or as in a case in rural India when a postman startled a tiger, used to seeing him on foot, by riding a bicycle. Occasionally tigers come to view people as prey. Such attacks are most common in areas where population growth, logging, and farming have put pressure on tiger habitats and reduced their wild prey. Most man-eating tigers are old, missing teeth, and unable to capture their preferred prey. For example, the Champawat Tiger, a tigress found in Nepal and then India, had two broken canines. She was responsible for an estimated 430 human deaths, the most attacks known to be perpetrated by a single wild animal, by the time she was shot in 1907 by Jim Corbett. According to Corbett, tiger attacks on humans are normally in daytime, when people are working outdoors and are not keeping watch. Early writings tend to describe man-eating tigers as cowardly because of their ambush tactics.

"Famous 'man-eater' at Calcutta—devoured 200 men, women and children before capture—India"

Man-eaters have been a particular problem in recent decades in India and Bangladesh, especially in Kumaon, Garhwal and the Sundarbans mangrove swamps of Bengal, where some healthy tigers have hunted humans. Because of rapid habitat loss attributed to climate change, tiger attacks have increased in the Sundarbans. The Sundarbans area had 129 human deaths from tigers from 1969 to 1971. In the 10 years prior to that period, about 100 attacks per year in the Sundarbans, with a high of around 430 in some years of the 1960s. Unusually, in some years in the Sundarbans, more humans are killed by tigers than vice versa. In 1972, India's production of honey and beeswax dropped by 50% when at least 29 people who gathered these materials were devoured. In 1986 in the Sundarbans, since tigers almost always attack from the rear, masks with human faces were worn on the back of the head, on the theory that tigers usually do not attack if seen by their prey. This decreased the number of attacks only temporarily. All other means to prevent attacks, such as providing more prey or using electrified human dummies, worked less well.

At least 27 people were killed or seriously injured by captive tigers in the United States from 1998 to 2001.

In some cases, rather than being predatory, tiger attacks on human seem to be territorial in nature. At least in one case, a tigress with cubs killed eight people entering her territory without consuming them at all.

In Captivity

In Ancient Roman times, tigers were kept in menageries and amphitheatres to be exhibited, trained and paraded, and were often provoked to fight humans and exotic beasts. Since the 17th century, tigers, being rare and ferocious, were sought after to keep at European castles as symbols of their owners' power. Tigers became central zoo and circus exhibits in the 18th century: a tiger could cost up to 4,000 francs in France (for comparison, a professor of the Beaux-Arts at Lyons earned only 3,000 francs a year), or up to $3,500 in the United States, where a lion cost no more than $1,000.

Tigers made to perform at Ringling Brothers and Barnum and Bailey Circus

China had over 4,000 captive tigers, of which 3,000 were held by about twenty larger facilities, with the rest held by some 200 smaller facilities. The USA had 2,884 tigers in 468 facilities. Nineteen states have banned private ownership of tigers, fifteen require a license, and sixteen states have no regulation. Genetic ancestry of 105 captive tigers from fourteen countries and regions showed that forty-nine animals belonged distinctly to five subspecies; fifty-two animals had mixed subspecies origins. As such, "many Siberian tigers in zoos today are actually the result of crosses with Bengal tigers."

The Tiger Species Survival Plan has condemned the breeding of white tigers, alleging they are of mixed ancestry and of unknown lineage. The genes responsible for white colouration are represented by 0.001% of the population. The disproportionate growth in numbers of white tigers points to inbreeding among homozygous recessive individuals. This would lead to inbreeding depression and loss of genetic variability.

Lion

The Lion is one of the largest, strongest and powerful felines in the world second only in size to the Siberian Tiger. They are the largest cats on the African continent and are unique among felines in a number of ways but the biggest difference between Lions and other cats is that they are incredibly sociable animals that live together in family groups known as prides. Lions are also part of the big cat family meaning that both males and females are able to roar. Despite having once roamed across much of Africa and even parts of Europe and Asia, the world's remaining Lion population now resides in sub-Saharan Africa. However, with Lion numbers thought to have dropped by 30% over the past 20 years these enormous predators are sadly becoming more and more vulnerable in their decreasing natural environment.

Anatomy and Appearance

Lions have a short coat of tawny or golden fur with a long tail that has a tuft of longer fur at the end. The markings on their coats are much fainter than the bold stripes and spots displayed on other felines which helps these large carnivores in going unseen when stalking prey in the long grasses. The Lion is one of the largest cats in the world with males being taller and heavier than females and displaying a mane of long hair around their faces (in fact, it is the only case in the feline world where males and females actually look different). Thought to be connected with testosterone levels, the mane of the male Lion ranges from blonde, to red, brown and black in color and covers their head, neck and chest. Lions have strong and powerful jaws that contain 30 teeth in total which includes four fang-like canines and four carnassial teeth that are perfectly designed for slicing through flesh.

Distribution and Habitat

Historically, Lions would have been found throughout much of Africa and even in parts of Europe and Asia as well. Today however, they have been pushed into more isolated pockets of their once vast natural range with the remaining African Lion population now only found in countries in sub-Saharan Africa. There is also still a small population of Asiatic Lions found inhabiting a remote part of the Gir Forest in India where there are an estimated 300 individuals remaining. Despite their dwindling numbers, Lions are actually incredibly adaptable animals that can and will inhabit very dry climates as they get most of the moisture they need from their food. They prefer areas of open woodland, scrub and long grasslands where there is not only plenty over cover but also a wide variety of prey. They are only not found in areas of rainforest or far into deserts.

Behavior and Lifestyle

Lions are unique among cats as they live together in strong social groups. A pride is made up of 5-15 related females and their cubs along with a generally single male (small groups of 2 or 3 though are not uncommon). Male Lions patrol a territory of around 100m² marking trees and rocks with urine and roaring to warn off intruders. Although male Lions can defend their pride to great effect, their position in the pride is constantly under threat from other males who try to take over their patch and if successful, they will kill any cubs that were sired by the previous male. Despite their enormous size, male Lions actually do hardly any of the hunting as they are often slower and more easily seen than their female counterparts. The Lionesses in the pride hunt together meaning that they are not only more successful on their trips, but they are also able to catch and kill animals that are both faster than them and much bigger.

Reproduction and Life Cycles

Both male and female Lions are able to reproduce between the ages of two and three but despite this, they will often not breed until the pride has been firmly established. After a gestation period that lasts for nearly four months female Lions give birth to between one and six cubs that are born blind and are incredibly vulnerable in their new surroundings. The fur of Lion cubs is covered in darker spots that help to camouflage them into their den to protect them whilst the adults have gone out to hunt. Sadly however, less than half of cubs make it to be a year old and four out of five have died by the time they are two, generally either from animal attacks or starvation. Remarkably though, the female Lions in the pride will have their cubs at around the same time and will help to suckle and care for the cubs of other females. Lion cubs suckle on milk until they are about six months old and although they won't begin actively hunting until they are about a year old, Lion cubs start to eat meat after 12 weeks or so.

Diet and Prey

The Lion is a large and carnivorous animal that survives only by eating other animals in order to sustain itself. Unlike other felines, Lions are not solitary hunters but instead the Lionesses work together in order to chase down and catch their prey with each female having a different strategic role. This strategy allows them to kill animals that are both faster and much larger than they are including Buffalo, Wildebeest and even Giraffe. Depending on the abundance and variety of prey species within their territory, Lions primarily catch Gazelle, Zebra and Warthog along with a number of Antelope species by following the herds across the open grasslands. They will not turn their noses up though at hunting alone should the situation arise and will happily steal the kill of another animal. Once the animal has been caught though, circumstances change as the females will allow the male Lion to eat first before indulging themselves. The cubs however are at the bottom of the pile and have to be content with what remains once the adults have finished.

Predators and Threats

The Lion is the most dominant predator within its environment meaning that other animals pose little or no threat to them, with the exception of Hyena packs that can cause fatal damages to Lions particularly when they are on their own and food is about. Lions are seen as a great threat by many other species including both Giraffes and Elephants which are easily capable of fatally injuring a Lion to try and warn it off. The biggest threat to Lions though is people who not only kill them out of fear (and historically as trophies), but have caused severe declines in Lion population numbers due to growing Human settlements and the clearing of land for agriculture. Lion numbers have also been severely affected by diseases passed through Hyenas from Wild Dogs, with more than 1,000 Lions having died from canine distemper between 1993 and 1997.

Interesting Facts and Features

Young Lion cubs spend a great deal of time playing together which actually helps them to develop their hunting techniques. This method of role-playing in cubs also helps females to determine whether or not they would be suited better to chasing and cornering prey, or catching and killing it. Although Lions are able to make a number of different sounds it is their roar that is most distinctive and not only serves to warn off strange males, but also allows members of the pride to find one another as its sound can travel up to six miles! The Lion has large paws with soft pads underneath and sharp retractable claws on the end of each toe which aid them in running, climbing and catching their prey along with also being good defence mechanisms. The structure of their feet and legs means that they are also able to jump distances over 10 meters.

Relationship with Humans

Lions have been admired and feared by people for centuries but due to both hunting and growing Human settlements, Lions have been wiped out from a vast portion of their historical natural range. Although they do not naturally see people as prey, African Lions have been known to sneak into villages (sometimes of great size) to find food, and are known to attack up to 700 people every year, with Lions being responsible for 100 human fatalities annually in Tanzania alone. In 1898, two Lions in Kenya (known as Tsavo lions - a species of mane-less Lion) became famous for killing and eating over 130 rail-road workers over a period of around 9 months. These powerful predators are largely feared by people who once hunted them as trophies and have also always done so out of fear.

Conservation Status and Life Today

Today, the Lion is listed by the IUCN as an animal that is Vulnerable from extinction in its natural environment in the near future. Although population numbers would have been vast in comparison to what they are now, over the past 20 years we have lost around 30% of the Lion population from sub-Saharan Africa with estimated population numbers ranging between 16,500 to 30,000 individuals (half of what they were in the 1950s). The main reason for the rapid demise of Africa's remaining Lion population is thought to be due to habitat loss as they are not only loosing big chunks of their native ranges, but are also often then forced onto farmland or into settlements to find food meaning they are often killed.

Jaguar

The jaguar (Panthera onca), is a New World mammal of the 'Felidae family'. It is one of four 'big cats' in the 'Panthera' genus, along with the tiger, lion and leopard of the Old World. The jaguar is the third largest feline after the tiger and the lion. The jaguar is the largest and most powerful feline in the Western Hemisphere.

The jaguars present range extends from Mexico (with occasional sightings in the southwestern United States) across much of Central America and south to Paraguay and northern Argentina.

Jaguar Characteristics

The jaguar is a compact and well-muscled animal. Rainforest jaguars are generally darker and considerably smaller than those found in open areas, possibly due to the fewer large herbivorous prey in forest areas. Jaguars grow to be about 1.62 – 1.83 meters (5.3 – 6 feet) in length and stand about 67 – 76 centimeters (27 – 30 inches) tall at the shoulders. Their tail is 2 – 3 feet (0.6 – 0.9 meters) long. Jaguars weigh around 36 kilograms (80 pounds). Larger jaguars have been recorded as weighing 131 – 151 kilograms (288 – 333 pounds).

The base coat of the jaguar is generally a tawny yellow in color, but can range to reddish-brown and black. The jaguar is covered in 'rosettes' for camouflage in its jungle habitat. The spots vary over individual coats and between individual Jaguars. The 'rosettes' may include one or several dots and the shape of the dots varies. The spots on the head and neck are generally solid, as are those on the tail, where they may merge to form a band. The underbelly, throat and outer surface of the legs and lower flanks are white.

Rare albino individuals, sometimes called 'white panthers', occur among jaguars, as with the other big cats.

The jaguar physically most closely resembles the leopard, although the jaguar is of sturdier build and its behavioral and habitat characteristics are closer to those of the tiger.

Jaguars are the second strongest of all mammals. The jaguar has a short, stocky limb structure which makes it well adaptable to climbing, crawling and swimming. Their heads are robust and their jaw is extremely powerful.

Jaguar Diet

Jaguars are carnivores (meat-eaters). Jaguars will eat a variety of animals including birds, eggs and mammals including capybaras, peccaries, tapirs, turtles and alligators. Jaguars often bury their prey after killing it, so that they can eat it later. The jaguar is often described as nocturnal, but is more specifically crepuscular (peak activity around dawn and dusk).

The jaguar is a largely solitary, stalk-and-ambush predator and is opportunistic in prey selection. It is also an apex predator (predators that, as adults, are not normally preyed upon in the wild in significant parts of their range by creatures not of their own species) and a keystone predator, playing an important role in stabilizing ecosystems and regulating the populations of prey species.

The jaguar has developed an exceptionally powerful bite, even compared to the other big cats. This allows it to pierce the shells of armored reptiles and to employ an unusual killing method. The jaguar bites directly through the skull of prey between the ears to deliver a fatal blow to the brain.

It has been reported that an individual jaguar can drag a 360 kilograms (800 pounds) bull 8 meters (25 feet) long in its jaws and crush the heaviest bones. The jaguar hunts wild animals weighing up to 300 kilograms (660 pounds) in dense jungle and its short and sturdy physique is therefore an adaptation to its prey and environment.

Jaguar Habitat

Jaguars are large, wild, graceful cats that live in rainforests, swamps, deserts and shrubby areas. These solitary felines often have dens in caves. It is strongly associated with the presence of water and is, along with the tiger, a very good swimmer.

Jaguar Behavior

Jaguars are territorial. Like most cats, the jaguar is solitary outside mother-cub groups. Adults generally meet only to court and mate and carve out large territories for themselves. Female territories, from 25 to 40 square kilometers in size, may overlap, but the animals generally avoid one another. Male ranges cover roughly twice as much area, varying in size with the availability of game and space and do not overlap. Scrape marks, urine and faeces are used to mark territory.

Like the other big cats, the jaguar is capable of roaring (the male more powerfully) and does so to warn territorial and mating competitors away. Intensive bouts of counter-calling between individuals have been observed in the wild. Their roar often resembles a repetitive cough and they may also vocalize mews and grunts. Mating fights between males occur, but are rare. Conflict is typically over territory. A males range may include that of two or three females however he will not tolerate intrusions by other adult males.

Jaguar Reproduction

Female jaguars reach sexual maturity at about 2 years of age and males at 3 or 4 years of age. The jaguar is believed to mate throughout the year in the wild, although births may increase when prey is plentiful. Female Jaguars estrous is 6 – 17 days out of a full 37 day cycle and females will advertise fertility with urinary scent marks and increased vocalization. Both male and female jaguars will range more widely than usual during courtship.

Mating pairs separate after mating and females provide all the parental care. The gestation period lasts 93 – 105 days. Female Jaguars give birth to up to 4 cubs but most commonly to only 2 cubs. The mother will not tolerate the presence of males after the birth of cubs, given a risk of infant cannibalism. This behavior is also found in the tiger.

The young are born blind, gaining sight after 2 weeks. Cubs are weaned at 3 months but remain in the birth den for 6 months before leaving to accompany their mother on hunts. They will continue in their mothers company for one to two years before leaving to establish a territory for themselves. Young males are at first leave, jostling with the older jaguars until they succeed in claiming a territory. Typical life span of the jaguar in the wild is estimated at around 12 – 15 years. In captivity, the jaguar lives up to 23 years, placing it among the longest living cats.

Jaguar Conservation Status

Jaguars are an endangered species due to loss of habitat and over-hunting by man and its numbers are declining.

All hunting of jaguars is prohibited in Argentina, Belize, Colombia, French Guiana, Honduras, Nicaragua, Panama, Paraguay, Suriname, the United States (where it is listed as endangered under the Endangered Species Act), Uruguay and Venezuela.

Hunting of jaguars is restricted to 'problem animals' in Brazil, Costa Rica, Guatemala, Mexico and Peru, while trophy hunting is still permitted in Bolivia. The species has no legal protection in Ecuador or Guyana.

Current conservation efforts often focus on educating ranch owners and promoting ecotourism (a form of tourism that appeals to the ecologically and socially conscious individuals). The jaguar is generally defined as an 'umbrella species' a species whose home range and habitat requirements are sufficiently broad that, if protected, numerous other species of smaller range will also be protected. Umbrella species serve as 'mobile links' at the landscape scale, in the jaguars case through predation. Conservation organizations may therefore focus on providing viable, connected habitat for the jaguar, with the knowledge that other species will also benefit.

Leopard

After tigers, lions and jaguars, the leopard is the fourth largest cat in the world. The most secretive and elusive of the large carnivores, the leopard is also the shrewdest. Pound for pound, it is the strongest climber of the larger cats and is capable of killing prey far larger than itself.

The coloring of the leopard varies from white to bright golden brown, spotted with black spots and rosettes. The rosettes consist of groups of 5 to 6 spots arranged in a tight ring.

The tail is longer than half the body length measured from head to tail. This fierce animal has small round ears and long whiskers growing from dark spots on the upper lip. The size of the leopard varies considerably. The leopard differs from the cheetah in having shorter legs, and rosette-like spots and is without the cheetah's black "tear" marks from eye to mouth.

The leopard ranges in size from 1 to almost 2 meters long, and weighs between 30 - 70 kg. Females are typically around two-thirds the size of males.

Habitat and Diet

Leopards easily adapt to a wide variety of habitats, altitudes, and temperatures, from high mountains to deserts to rain forests. The largest leopards are named after the regions where they're found, like the North African leopard and the Persian leopard. All can live without drinking water for as long as 10 days, getting the moisture they need from their food. All they need is some brush, thick vegetation, caves, or rocky terrain for hiding and hunting. Leopards usually rest during the heat of the day in bushes, rocks, caves, or even up in a tree, depending upon their habitat. Of the large cats, leopards are the most arboreal; they have long tails to help them to balance on narrow tree branches.

Unlike most cats, leopards are strong swimmers and are one of the few cats that like water, although they are not as aquatic as tigers. They are great athletes, able to run in bursts up to 36 miles an hour (58 kilometers per hour), leap 20 feet (6 meters) forward in a single bound, and jump ten feet (3 meters) straight up.

Leopards have incredible strength and can climb as high as 50 feet (15 meters) up a favorite tree while holding a fresh kill in its mouth, even one larger and heavier than themselves! They stash food up high so other predators such as lions or hyenas can't get it. This way, leopards can return to eat more at a later time. One leopard was spotted dragging a 220-pound (100 kilograms) young giraffe into heavy brush to hide it. Leopards are usually nocturnal, resting by day and hunting at night.

Stories from both Africa and Asia tell of the leopard's ability to enter a village and snatch a sleeping dog without being detected. The leopard is a champion hunter and has a variety of stealth attacks that catch its prey off guard. From dropping on prey out of trees to stalking prey at waterholes or in dry grasses, often slinking along on the belly, a leopard doesn't have a predictable pattern to hunting.

The cat uses its vision, keen hearing, and whiskers, rather than its sense of smell, while hunting. The whiskers face forward when the leopard is walking and move back when the cat is sniffing; they stick out sideways when the leopard is resting. The big cat stalks and pounces rather than chases its prey over long distances. It grabs or swats prey, using retractable claws. Prey is killed with a bite to the throat. Leopards are carnivores and eat any meat item they can find: monkeys, baboons, rodents, snakes, amphibians, large birds, fish, antelope, cheetah cubs, warthogs, and porcupines.

Family Life

Adult males and females find each other through scent, and a male may follow a female around for several days before she is ready to breed. Even then, a breeding encounter can be dangerous, as both cats have sharp claws and teeth and know how to use them.

A pregnant female uses a cave, burrow, or an opening in thick vegetation as a birthing den. She stays with her cubs, usually two or three in a litter, continually for their first few days before she finally ventures out for food. The cubs are born with little hair and their eyes are sealed. As they get older, Mom may have to leave her cubs for several days while she hunts. At about three months of age, the cubs venture out with their mother to learn hunting skills.

Like all young cats, leopard cubs like to play "stalk, pounce, and chase." Have you ever seen a house cat creep slowly after a bird or mouse? That's stalking. A quick leap and a grab with the claws is a pounce, and the chase comes if the prey gets away. Leopard cubs play by practicing these behaviors on their siblings and even on their mother. It's a good way to learn how to survive when they get older.

Young leopards are usually ready to head out on their own between 12 and 18 months and are ready to start a family of their own at two to three years.

Leopards rarely roar; their voice is more of a raspy bark. White spots on the tip of the tail and back of the ears help leopards locate and communicate with each other in tall grass. Except during breeding season, leopards lead a solitary life, marking their territory with urine, rubbing the face against rocks or tree trunks, scraping at the ground or rolling on it, and by shredding tree bark with their claws. Fighting between males is common and can be deadly.

Conservation

Although the leopard is an adaptable cat, able to live in various habitats, some leopard subspecies are at critical risk. Leopard-skin coats were legal for many years and are still sold secretly.

Monitoring the wild populations of Amur leopards is critical to understanding population trends, which will determine the effectiveness of current conservation measures. The Amur Leopard and Tiger Alliance and its partner, the Wildlife Conservancy Society, have been monitoring the leopards since 1997. The cats (and other wildlife) have been surveyed in winter by counting tracks in the snow along established animal paths to estimate leopard density and population size. In 2002, camera traps were added to allow researchers to identify individual leopards by their unique spot patterns and monitor individual animals over many years. Used year-round, the images uncover the secret lives of these cats walking, stalking, and rolling playfully on the ground. Camera images also captured prey species including sika deer, boars, and yellow-throated martens.

The Amur leopard is under siege from a variety of pressures including poaching of the leopards and their prey, loss of habitat due to forest fires, inbreeding due to tiny, isolated populations, human development and activities in their habitat, and lack of political commitment to conservation.

Leopards and Humans

Benin water vassel shaped in the form of a leopard

Leopards have been known to humans throughout history, and have featured in the art, mythology, and folklore of many countries where they have historically occurred, such as ancient Greece, Persia, and Rome, as well as some where they have not existed for several millennia, such as England. The modern use of the leopard as an emblem for sport or a coat of arms is much more restricted to Africa, though numerous products worldwide have used the name. During the Benin Empire, the leopard was commonly represented on engravings and sculptures and was used to symbolise the power of the king or *oba*; since the leopard was considered the king of the forest. Leopards were also kept and paraded as mascots, totems and sacrifices to deities. As a result of their association with kings in Africa, the leopard's pelt is often seen today as a symbol of aristocratic rank, chiefs using it as a part of their traditional regalia.

Three leopards (loggerheads) on the flag of Shropshire, England.

The lion passant guardant or leopard is a frequently used charge in heraldry, most commonly appearing in groups of three. The heraldic leopard lacks spots and sports a mane, making it visually almost identical to the heraldic lion, and the two are often used interchangeably. These traditional lions passant guardant appear in the Coat of arms of Dalmatia, coat of arms of England and many of its former colonies; more modern naturalistic (leopard-like) depictions appear on the coat of arms of several African nations including Benin, Malawi, Somalia, the Democratic Republic of the Congo and Gabon, which uses a black panther.

Leopard domestication has also been recorded—several leopards were kept in a menagerie established by King John at the Tower of London in the 13th century; around 1235, three of these animals were given to Henry III by Holy Roman Emperor Frederick II.

Tourism

A female leopard in the Sabi Sands of South Africa near a game vehicle

In protected areas of several countries, wildlife touring programs and safari ventures offer sightings of leopards in their natural habitat. While luxury establishments may boast the fact that wild animals can be seen at close range on a daily basis, the leopard's camouflage and propensity to hide and stalk prey typically make leopard sightings rare. In Sri Lanka's Yala National Park, leopards have been ranked by visitors to be among the least visible of all animals in the park despite their high concentration in the reserve.

In South Africa, safaris are offered in numerous nature reserves such as the Sabi Sand Game Reserve. In Sri Lanka, wildlife tours are available in the Yala and Wilpattu National Parks. In India, safaris are offered in the Madhya Pradesh and Uttarakhand national parks as well as in the Pali district of western Rajasthan.

Man-eating

The Panar Leopard, shot by Jim Corbett in 1910 after allegedly killing 400 people

Most leopards avoid people, but humans may occasionally be targeted as prey. Most healthy leopards prefer wild prey to humans, but injured, sickly, or struggling cats or those with a shortage of regular prey may resort to hunting humans and become habituated to it. Although usually slightly smaller than a human, an adult leopard is much more powerful and easily capable of killing one. Two extreme cases occurred in India: the first leopard, "the Leopard of Rudraprayag", killed more than 125 people; the second, the "Panar Leopard", was believed to have killed more than 400. Both were killed by the renowned hunter and conservationist Jim Corbett. Man-eating leopards are considered bold and difficult to track by feline standards and may enter human settlements for prey, more so than lions and tigers.

Cheetah

The Cheetah is a large and powerful feline that was once found throughout Africa and Asia and even in parts of Europe. Today however, it is found in only a few remote regions of its once vast natural range, primarily due to growing Human settlements and the hunting of them for their fur. There are widely considered to be five different subspecies of Cheetah that vary only very slightly in coloration and are most easily distinguished by their geographic location. Although they are not considered to be part of the 'big cat' family as they cannot roar, Cheetahs are one of Africa's most powerful predators and are most renowned for their immense speed when in a chase. Capable of reaching speeds of more than 60mph for short periods of time, the Cheetah is the fastest land mammal in the world.

Anatomy and Appearance

The Cheetah has a long and slender body that is covered in coarse yellowish fur and dotted with small black spots. Its long tail helps with balance and changing direction quickly and unlike the rest of the Cheetah's body, there are ringed markings along the tail which ends in a black tip. Cheetahs have small heads with high set eyes that aid them when surveying the surrounding grasslands for potential prey. They also have distinctive black "tear marks" that run from the inner eye, along their nose and down to the outside of their mouths, which are thought to help protect them from being blinded by the bright sun. The exceptional speed of the Cheetah is caused by a number of things including having strong and powerful hind legs, and an incredibly flexible and muscular spine which allows the Cheetah to not only sprint quickly but also makes them very agile. They also have non-retractable claws which dig into the ground, giving the Cheetah better grip at high speed.

Distribution and Habitat

The Cheetah once had a vast historical range that stretched across a number of continents, but their distribution today is much more scattered with a small number found in Iran and the majority found in sub-Saharan Africa. Although Cheetahs are still found in a few different parts of eastern and southern Africa, the highest population of wild Cheetahs is now found in Namibia in south-western Africa. Cheetahs are most commonly found stalking prey on the vast, open grasslands but they are also found in a variety of other habitats as well including deserts, dense vegetation and mountainous terrain, providing that there are both adequate supplies of food and water. Cheetahs are one of Africa's most vulnerable felines with population numbers being mainly affected by growing Human settlements that encroach on their native habitats.

Behavior and Lifestyle

The Cheetah is unique amongst Africa's felines primarily because they are most active during the day, which avoids competition for food from other large predators like Lions and Hyenas that hunt during the cooler night. They are also one of the more sociable Cat species with males often roaming in small groups, generally with their siblings, and oddly enough, it is the females that are more solitary animals apart from the 18 months or so that they spend looking after their cubs. Cheetahs are fiercely territorial animals that patrol large home ranges and often overlap those of other Cheetahs, and indeed Lions, with females tending to roam across a much larger range than males. They are typically shy and very stealthy animals so that they are able to hunt for prey in the hot daylight hours without being spotted so easily.

Reproduction and Life Cycles

After a gestation period that lasts for around 3 months, the female Cheetah gives birth to between two and five cubs that are born blind and incredibly vulnerable in the African wilderness. The cubs suckle from their mother for the first few months when they begin eating meat, and start to accompany her on hunting trips as they are able to learn how to hunt from watching her. Cheetah cubs learn the majority of their hunting techniques through playing with their siblings, and remain with their mother until they can hunt successfully and leave to find their own territory at between 18 months and 2 years old. Sadly, one of the main reasons for such drastic declines in Cheetah numbers is that up to 75% of Cheetah cubs don't live to be older than 3 months, as their mother has to leave them every day to find food to feed them, leaving the vulnerable cubs helpless in the face of predators.

Diet and Prey

The Cheetah has exceptional eyesight and so hunts using sight by first stalking its prey (from between 10 to 30 meters away), and then chasing it when the time is right. Cheetahs often kill their prey in vast open spaces but tend to drag it into a hiding place to prevent it from being scavenged by other animals. The Cheetah has to do this as it cannot eat its prey straight away, due to the fact that they are immensely hot after the chase and need time to cool down before they can feast. Cheetahs are carnivorous animals meaning that they only hunt and kill other animals in order to get the nutrition that they need to survive. They mainly hunt large herbivores including Gazelle and a number of large Antelope species such as Wildebeest, along with Zebras and smaller mammals like Hares. The exact diet of the Cheetah though tends to depend on its location.

Predators and Threats

The adult Cheetah is a dominant predator in its environment and is therefore not seen as prey (more as competition) by other large predators. The Cheetah cubs however, are incredibly vulnerable particularly when their mother is off hunting, and are preyed upon by a number of animals including Lions and Hyenas, but also large avian species such as Eagles and Vultures. The biggest threat to Cheetahs is people, who have not only taken vast areas of its natural habitat for themselves, but have also turned large regions of land into National Parks. Although these areas seem to be showing increases in both Lion and Hyena populations, Cheetah numbers are much lower as there is more competition for food with the increasing numbers of these other large predators.

Interesting Facts and Features

One of the most distinctive differences between the different subspecies is that their fur color changes slightly, depending on their surrounding environment. Those Cheetahs found in the more arid desert regions tend to be lighter and have smaller spots than the King Cheetahs that roam the south African grasslands, and are darker, slightly bigger and have larger spots. Although Cheetahs cannot roar, they do make a variety of different sounds including a high-pitched yelp which can be heard for more than a mile away. The Cheetah is one of the most distinctive of the world's felines and although it is fairly commonly confused with the Leopard, its name is thought to have come from the Hindu word 'chita' which means the 'spotted one'.

Relationship with Humans

Despite their fierce appearance, Cheetahs have actually been domesticated by local people for thousands of years, as they were used to help with hunting for food for the villagers. They have also been kept in captivity for a long time, but as they do not seem to produce particularly healthy individuals, wild Cheetahs were regularly caught to restore the bloodline. Hunted by people as trophies throughout much of their natural range, Cheetahs have completely disappeared from numerous places and along with habitat loss to growing Human settlements and the clearing of land for agriculture, Cheetah numbers are still declining rapidly in many areas today.

Conservation Status and Life Today

Today, the Cheetah has been listed by the IUCN as a species that is Vulnerable from extinction in its natural environment in the near future. Habitat loss along with the increase of natural parks that are home to a large number of competing predators has led to drastic declines in the world's Cheetah population. There are now estimated to be between 7,000 and 10,000 individuals left in the wild, along with a growing number being found in zoos and animal institutes all over the world.

Puma

Puma, (Puma concolor) is also called mountain lion, cougar, panther (eastern U.S.), or catamount (archaic) large brownish new world cat comparable in size to the jaguar—the only other large cat of the western hemisphere. The Puma is a member of the family Felidae, has the widest distribution of any new world mammal, with a range extending from southeastern Alaska to southern Argentina and Chile. Pumas live in a variety of habitats, including desert scrub, chaparral, swamps, and forests, but they avoid agricultural areas, flatlands, and other habitats lacking cover (vegetative or topographic). Six subspecies of Puma concolor are recognized by most classifications.

Puma (Puma concolor) cubs and mother.

Pumas living near the Equator are generally smaller than those living farther north and south. Males in North America average 62 kg (136 pounds), but rare individuals can exceed 100 kg; length is about 1.2 meters (4 feet), excluding the 0.75-metre (2.5-foot) tail. Females are somewhat shorter and average about 42 kg. The specific name concolor ("of one color") refers to the puma's fur, which is uniformly brown on the back, sides, limbs, and tail. (The name puma is a native Peruvian term.) The shade of brown varies geographically and seasonally from gray to reddish brown, and some black pumas have been reported; facial color patterns are also variable. The underside is lighter. The long tail is commonly tipped with black and usually held close to the ground when the puma is walking.

Population Status and Taxonomy

At the time of European settlement in the New World, pumas occupied all of what are today the lower 48 United States and southern Canada. At present they are primarily found west of 100° W longitude (approximately central Texas to Saskatchewan) except for southern Texas and are an endangered population in Florida (the Florida panther, P. concolor coryi). Information is lacking for Central and South America, although most suitable habitats there are thought to be inhabited. A subspecies known as the eastern cougar (P. concolor cougar), which once inhabited the eastern United States and southern Ontario and was listed as endangered in 1973, was declared extinct in 2011.

Florida panther (Puma concolor cougar).

Since 1950 pumas have been eliminated from the states of Arkansas, Louisiana, Tennessee, and West Virginia. Until the 1960s pumas were taken by U.S. government hunters and subject to state bounties. Since 1970 control efforts in the United States and Canada have focused on known livestock killers, and most states and provinces now manage populations for sustained sport hunting. In most of the western United States and Canada, populations of mountain lions (P. concolor concolor) are thought to be stable or increasing except where habitat is being fragmented by urban sprawl.

Although cougars are elusive and usually avoid people, there are about four attacks and one fatality per year on humans in the United States and Canada. Most victims are children or adults traveling alone. Risk can be minimized by walking in groups and keeping children within sight. An aggressive human response can avert an impending attack and can repel an attack in progress.

The power and stealth of the puma have come to epitomize the wilderness, and the cat has therefore received prominent consideration in conservation and recovery efforts. For example, habitat corridors are planned between large natural areas in order to benefit large carnivores such as the puma. Research has demonstrated that dispersing pumas readily find and use habitat corridors, and radio tracking of these wide-ranging predators can be used to identify appropriate areas to conserve as corridors.

Habitat and Habits

Pumas can adapt to a wide variety of climates and habitats. They are found throughout Central and South America. In North America, however, hunting has reduced their range to isolated areas in Mexico, western U.S. wilderness areas, southern Florida and southwestern Canada.

Unlike other cats, pumas do not live in packs. They live by themselves in large territories, which they mark using urine or feces. Pumas are solitary creatures for good reason. They need hunting room to find enough food to live. Only a few pumas can live in a 30-square-mile (78 square kilometers) radius.

Pumas are crepuscular creatures. They hunt during the twilight hours of dawn and dusk. The low light hides them from their prey. Their excellent night vision gives these predators a distinct advantage during these times of day.

Diet

Pumas hunt using stealth and strength. To kill their prey — usually deer, elk or moose — the puma sneaks up behind the other animal and then chomps down on the prey's neck with its powerful jaws. To make the kill last longer, it hides its prey from other animals and eat bits of the carcass for several days. A puma typically kills a deer every 10 to 14 days.

Pumas can drag prey several times their weight and, given the opportunity, will hunt domestic animals, such as sheep, pigs, horses and other livestock. They will also eat small animals such as porcupines, coyotes, rabbits, armadillos, capybaras, squirrels and raccoons. Some pumas hunt more than others. Patagonian pumas kill around 50 percent more prey than their North American counterparts.

Offspring

After mating, the female will carry her young for a gestation period of 84 to 106 days. Then she will give birth to a litter of one to six cubs. These cubs are typically covered with spots that can help them blend in with their surroundings. A female will have a litter once every two years.

Classification/Taxonomy

According to the Integrated Taxonomic Information System (ITIS), the taxonomy of pumas is:

- Kingdom: Animalia
- Phylum: Chordata
- Class: Mammalia
- Order: Carnivora
- Family: Felidae
- Genus: Puma (changed in 1993 from Felis)
- Species: Concolor
- Subspecies: Six subspecies are recognized: Puma concolor anthonyi, Puma concolor cabrerae, Puma concolor concolor, Puma concolor costaricensis (Costa Rican puma), Puma concolor couguar (Eastern puma) and Puma concolor puma.

Conservation Status

Pumas once ranged from the Pacific to the Atlantic, but they were eliminated from eastern North America within 200 years of colonization. During the 20th century, large numbers of pumas were killed in the Midwest and Eastern United States by farmers and ranchers who didn't like the puma stealing their livestock. As a result, pumas almost became extinct in those areas.

The puma population is not increasing and more are being killed now than ever before because of habitat destruction, trophy hunting, poaching, poisoning and being struck by vehicles on roadways ("road kill").

Pumas are listed as "Least Concern" for extinction because they are so widespread, according to the IUCN. However, they are considered to be declining, and the Florida panther is considered endangered, and in Brazil, pumas are considered "Near Threatened."

The U.S. Fish and Wildlife Service lists the Florida panther, the Costa Rican puma and the Eastern puma as endangered.

Pumas are a protected species in many areas, with hunting prohibited in many South American nations. In the United States, hunting is legal in many Western states. California, however, banned hunting in a referendum in 1990. Hunting regulations are in place in Canada, Mexico and Peru.

Bear

Bears are large dog like mammals found all around the world. There are eight different species of bear that are found in a wide range of habitats in both the Northern and Southern Hemispheres, mainly the Americas, Europe and Asia.

Most bears are nocturnal, solitary animals only really congregating during the bears mating season. The mother bear will then raise her cubs until they too, are old enough to live on their own. Bears generally have an excellent sense of smell and are also fantastic at climbing trees, swimming and are able to run at speeds of up to 35 mph for short periods of time.

All bears are generally classified as carnivores but most species of bear today have adapted various herbivorous traits. For example, the giant panda has a diet that primarily comprises of bamboo. Most species of bear tend to hunt fish over land animals

The ability of early bears to adapt to a variety of living conditions and major environmental changes, accounts for why they have been able to be so diverse, which also lead to the development of the different species of bears. Each one changed in ways that allowed them to be able to live in a given area, developing the adaptations to survive the specific conditions.

For example, some adaptations include the coloring of their fur for blending into their surroundings, its thickness to resist the weather conditions and some others. The type of food that they eat and where they get it from also is geographically diverse. While some learnt to catch salmon fish, others developed long tongues to catch termites.

Bears seem to have a very bad reputation out there in society because their aggressiveness, but this mainly happens when they are threatened. Bears are solitary animals that are often misunderstood. They generally will mind their own business out there in the wild, but can get upset when they feel themselves or their young threatened.

Today people continue to move closer and closer to the natural environment of bears. Then they complain when they get into their trash or picnic areas. We have to remember that how they behave is due to their natural instincts.

There are only eight living species of bears that are found throughout the world. Most of them are either black or brown in color. The one that stands out is the Polar Bear who is white. This is due to the fact that they live in the very cold regions with lots of snow and the white coloring helps them to easily blend into their surroundings.

The eight living bear species are:

- Giant Panda Bear (Ailuropoda melanoleuca)
- Spectacled Bear (Tremarctos ornatus)
- Sun Bear (Helarctos malayanus)
- Sloth Bear (Melursus ursinus)
- Asian Black Bear (Ursus thibetanus)
- American Black Bear (Ursus Americanus)
- Brown Bear (Ursus arctos)
- Polar Bear (Ursus maritimus)

Bears are well known for their size and powerful strength. They are often the winners in the wild when there is a battle among them and another animal. The warning signs of a bear not being happy include standing up tall and roaring. It is said that you should make yourself appear as large as possible when a bear is in such a stance. Chances are they will run away from you instead of attacking.

It is the fear of being attacked that has caused and unnatural fear of bears throughout our society. Attacks can occur but they are very infrequent. Yet those occurrences get sensationalized and that gives people a very bad impression about them. In North America, there are registered two fatal bear attacks of Black bears during 2011 and two more from Brown Bears.

Most bears are content to live in the wild and to roam around looking for food. They consume a variety of foods including nuts and berries. They can be found in polar ice, grasslands or forests. They can also be found on the open plains or in heavily wooded areas.

Bears have no problem to adjusting in captivity. In fact, they are among the most popular attractions at zoos all over the world.

There are several species of bears that have already become extinct. This is due to slow reproduction rates and the destruction of their environment. There are some other species of bears that are also endangered or vulnerable. Even with active groups doing all they can for awareness purposes and to help get their numbers up there is the risk.

Gray Wolf

Gray wolves, also known as "timber wolves" or "common wolves," are the largest members of the wild canid family. The average measures between 3.5 and 5.5 feet long, and weighs approximately 80 pounds. Despite their name, a wolf's fur can range in color from almost pure white to black. The most common shade is tawny brown.

Gray wolves are social animals, and form packs of up to 12 members, usually consisting of a nuclear family. Wolf packs contain a dominant, or "Alpha" male and female, their offspring, and other wolves not related to the dominant pair. All of the wolves have a rank that determines their roles and hierarchy in the pack. Wolves are well known for their distinct sound: the howl. This unique noise can travel for several miles. Wolves howl for a number of reasons, including communication with other packs and with their own pack over long distances, to gather a pack for a hunt, to warn off intruders, or to attract mates.

Wolf mating season is from January to February and is usually limited to the alpha pair, although males will sometimes mate with subordinate females as well. The alpha female will have litters of up to 6 pups each year. During their first months, pups are completely dependent on their mother's milk. They are gradually weaned so that when they are seven months old, they are ready to start hunting with the adults. Raising the wolf pups is a responsibility shared by all members of the pack, and young wolves will typically stay in their pack for two to three years. Eventually, the young wolves will leave to join other packs, or find a mate and form their own packs. A wolf's average life span is 6-8 years, although the record wolf life span is around 20 years.

Gray wolves once had one of the largest ranges of any species, surpassed only by humans. It spanned throughout Eurasia, North America, and Africa, covering the entire continental northern hemisphere. Currently, the gray wolf inhabits a reduced portion of its range, largely due to human activity and habitat destruction.

Habitat

Wolves can thrive in a variety of environments from the tundra to the woods, forests, grasslands, or even deserts. Wolves require large areas capable of supporting stable populations of their prey. The range of a wolf pack typically varies from 80 to 300 miles, depending on the number of prey species inhabiting the region. In extreme northern regions, territories can cover over 1,000 square miles.

Diet

Wolves primarily use their hunting energy to target large hoofed animals like deer, moose, bison, mountain goats, sheep, elk, and caribou. Despite the wolves' extremely acute hunting skills, most of the animals they stalk eventually escape. Wolves are also opportunistic hunters, and will target small prey if it is available to them, such as rabbits, snakes, mice, birds, fish, and beavers.

Wolves can go longer than a week without eating, with one known case of a wolf surviving for 17 days without food.

Wolves have 42 highly specialized teeth. Their canines, or "fangs," measure around 2-1/2" long, and they also have large molars for crushing and grinding bones. The jaws of a wolf are extremely powerful, and the wolf has a biting capacity of between 1,000 and 1,500 pounds per square inch.

Giant Otter

The Giant Otter features a very long body, similar to that of a weasel. Many people find this particular Otter to be different in size and shape from others. For that reason they are often mistaken for another type of animal.

Distribution

Throughout South America is where you will find the Giant Otter. They prefer to reside in the rivers and streams that hold freshwater. It is also common for these regions to flood during the rainy season.

Behavior

They are very social which is why you will find them in very large groups. It is also known as the noisiest of all species due to the constant communications that go on among them. It is very fascinating though to listen to all of the activity going on.

You will find that they are very active during the day. Then they retreat to their dens on land as the sun goes down. These Otters prefer to live in groups that can have up to 20 members in them. They have a complex hierarchy that takes place within these groups. For example the females with pups take precedence over all the rest.

They are very protective of the members in their groups too. Usually it is the males that take care for defending them but instances of females doing the battles have been recorded as well.

Diet/Feeding

The Giant Offer primarily feeds on various types of fish in the water. They also enjoy feeding on crab and small snakes.

Reproduction

The reproduction of the Giant Otter leaves a great deal to be figured out. Experts can only assume what really goes on in the wild. They have observed such efforts in captivity but feel that there will also be some differences based upon the location. As long as there is plenty for the Otters to survive on they will take part in mating all year long.

The males are the ones that will initiate the process though. About 70 days after conception the pup will be born. A female may be ready to mate again in a couple of months but it can be up to 2 years before she does.

They are good caregivers, but research shows that stress can cause them not to create enough milk or to abandon their pup. This is both in captivity and in the wild. The average lifespan for a Giant Otter is 8 years in the wild and 17 in captivity.

Conservation

The future for the Giant Otter has been looking grim for many decades. However, with conservation efforts in place there may still be some hope for them after all. There are less than 5,000 of them in the wild and approximately 60 of them in captivity. They do seem to adapt well to changes in their natural environment though which is a plus.

Right now it is categorized as being endangered. There continues to be plenty of research too about introducing them to new environments as a means of increasing their numbers. Poaching continues to be a huge problem which is the illegal hunting of these animals.

Commercial fishermen are being asked to use nets that won't trap or kill the Giant Otter. They are more expensive though than what the other nets cost so many of them still haven't made the transition.

Human Interaction

Humans are the only real predators that the Giant Otter has. However, there have been some reports of cougars and jaguars killing them in certain areas. This is very likely due to changes in the natural habitat for those types of animals though.

White-tailed Mongoose

The white-tailed mongoose is the largest of the world's mongoose species and is common across the continent.

The white-tailed mongoose's body, from back to front, starts with uniquely white, long and bushy tail. From there its fur fades darker towards its hind legs which are just as black as its front ones. Its torso and back are both a grey color with a lighter, warmer orange-brown undercoat shining through. From the front its light face, pointy nose and eyes are the only things sticking out from within vegetation when a spotlight is put on this creature during the night as it camouflages. It is one of the larger species of mongoose found within the Kruger, with the average weight of males at around 3.6kg (lb), with females minutely smaller at a mean of 3.4kg (lb).

Distribution

The White-tailed Mongoose, unlike the mongoose species already discussed, can in fact be found in parts of the Sahara Desert, but only near the Nile in parts of Sudan right up to the border with Egypt. Its range stretches from the southernmost parts of KwaZulu-Natal in South Africa up along the coast towards Somalia and Kenya as well as parts off the coast of Sudan and more inland in areas of Zimbabwe, northern Namibia, Angola, Tanzania, Kenya, Malawi, the Central African Republic, Cameroon and into Western Africa. This range does however exclude the densest parts of the rainforest, the Namib Desert and most of the old Cape Province of South Africa.

Status

After a recent reassessment of previously collected data, environmentalists have concluded that the conservation status of this species, like all other mongoose species found in the Kruger National Park, is nothing to be concerned about. Their broad range across Africa, far and wide, together with their substantial populations in protected areas has made these animals a very stable group going forward. Altitude doesn't even seem to bother them which is evident when looking at populations in Ethiopia who live in altitudes of up to 3500m above sea level in savanna and woodland areas, living as strict insectivores most of the time, a very stable source of food less affected by human populations.

Habitat

Like other mongooses, White-tailed mongooses live where they can make a decent living. Savannah areas, woodland areas and even grasslands are all home to some population of white-tailed mongoose or another, keeping in mind it should be lower than 3500 meters above sea level, and be below the tree line. Wetter, more moist and humid areas aren't seen as hospitable to these creatures, and this is why they aren't found in rainforests or high rainfall tropical areas. Semi-desert and desert neighborhoods are also out of the picture when going house hunting with these animals, mainly due to the lack of an adequate food and drinking water supply.

Social Organization

An average white-tailed mongoose will sleep by its lonesome self during the day, while hunting and foraging for food in the same fashion at night. Only an estimated 13% of the total population of these animals are known to live in groups with other of their kind and share social interactions with them. They are territorial and also scent mark through the use of many different glands like fellow mongooses have been shown to do, with a particular preference for anal gland marking, most probably because of its dominance in the potency category.

Social Behavior

Their social time is preoccupied with foraging for the most part, with close to their whole day devoted to this task. The do however perform certain odd movements and exercises during this time while walking from one foraging ground to the next. They have been known to walk in a zig zag pattern, stop every now and again to lick and bite air for no particular reason and also make a muttering sound when harvesting insects that need to be dug out from under rocks or in holes. They walk throughout their whole home range during one eve's foraging, and try and scent mark as frequently as possible as they go along.

Reproduction

White-tailed Mongooses have seldom been seen or recorded mating, and this is why there is little known about their courtship except that they mate many times within half an hour, where the female stops copulation each time. This is thought to occur during the dry season, just in time for their litter of 1 to 3 young to be born during the wet season between February and May, after a tough period of gestation for their mother. They mate only once annually compared to the 4 times a year some other mongoose species are able to reproduce and give birth to new young.

Anti-predator Behavior

When attacked, at first these animals don't seem like they could harm anyone. Their initial reaction is to freeze and potentially start running for cover faster than you might think a mongoose can run, very harmlessly. It is when their attackers catch up with them that their true colors start showing. Immediately they start their very well-rehearsed intimidation act where they erect all the hair on their back, along with their tails, to make them seem bigger than they really are. What follows is a fowl smelling wave of heavy anal sac secreting. If push comes to shove these animals may even engage in a fight with their predators.

References

- Turner, A. and Antón, M. (1997). The Big Cats and Their Fossil Relatives: An Illustrated Guide to Their Evolution and Natural History. Columbia University Press. ISBN 978-0-231-10228-5
- Mazák, J.H. and Groves, C.P. (2006). "A taxonomic revision of the tigers (Panthera tigris) of Southeast Asia". Mammalian Biology, Zeitschrift für Säugetierkunde. 71 (5): 268–287. doi:10.1016/j.mambio.2006.02.007
- Gray-wolf-facts-information, learn-about-wildlife: americanexpedition.us, Retrieved 21 May 2018
- Actman, Jani (24 February 2017). "Cat Experts: Ligers and Other Designer Hybrids Pointless and Unethical". National Geogrpahic.com. Retrieved 27 August 2018
- Karanth, K. Ullas & Sunquist, Melvin E. (1995). "Prey Selection by Tiger, Leopard and Dhole in Tropical Forests". Journal of Animal Ecology. 64 (4): 439–450. doi:10.2307/5647. JSTOR 5647

- Puma-mammal-species, animal: britannica.com, Retrieved 11 March 2018
- Wozencraft, W.C. (2005). "Order Carnivora". In Wilson, D.E.; Reeder, D.M. Mammal Species of the World: A Taxonomic and Geographic Reference (3rd ed.). Johns Hopkins University Press. p. 546. ISBN 978-0-8018-8221-0. OCLC 62265494
- Smith, J. L. D.; Sunquist, M. E.; Tamang, K. M.; Rai, P. B. (1983). "A technique for capturing and immobilizing tigers". The Journal of Wildlife Management. 47 (1): 255–259. doi:10.2307/3808080. JSTOR 3808080
- Wildlife-leopard: sa-venues.com, Retrieved 10 May 2018
- Kupferschmidt, K. (2015). "Controversial study claims there are only two types of tiger". American Association for the Advancement of Science. doi:10.1126/science.aac6905. Retrieved 27 June 2015

Chapter 3

Ungulate Mammals

Ungulates are large mammals, such as rhinoceroses, horses, pigs, giraffes, cattle, etc. These can be classified as odd-toed and even-toed ungulates. This chapter closely discusses about ungulates and their different types to provide an extensive understanding of ungulate mammals.

Ungulate

Ungulates are a group of large mammals that are distinguished from other animals by the presence of hooves. They are an extremely well-known and economically important group that include animals such as horses, camels, cows, sheep, goats, deer, pigs, giraffes, hippos, rhinos and many more.

Ungulate species are separated into two orders: Perissodactyla and Artiodactyla. The Perissodactyls are commonly known as the odd-toed ungulates and includes horses, rhinos and tapirs. The Artiodactyla consists of the even-toed ungulates and is a much larger order with around 220 different species. Cows, sheep, goats, buffalos, deer, giraffes, pigs and camels are just a few of the many even-toed ungulates that currently exist on Earth.

Almost all ungulates are herbivores and they have evolved ways to break down difficult compounds in plants that would be toxic to other mammals such as humans.

Even-toed Ungulate

Even-toed ungulate is the common name for any of the hoofed, mostly herbivorous, terrestrial mammals comprising the order Artiodactyla, characterized by a double-pulley astragalus (a bone in the ankle joint) and an even number of functional toes (two or four), with the main limb axis passing between the middle two digits. Artiodactyls include such well-known members as pigs, peccaries, hippopotamuses, camels, chevrotains (mouse deers), deer, giraffes, pronghorns, antelopes, sheep, goats, and cattle. Artiodactyla is one of two living orders of ungulates (hoofed mammals), the other being Perissodactyla, the odd-toed ungulates.

With more than 220 species, artiodactyls are the most abundant large land mammals living today. They also are of great nutritional, economic, and cultural importance to humans, and in particular the domestic species are highly sought for such functions as food, fur, milk, fertilizer, medicinal products, and as draught animals. And wild species such as antelopes and deer provide both food and sports hunting, in addition to adding to the wonder of nature. Ecologically, the wild artiodactyls provide important roles in terrestrial food chains.

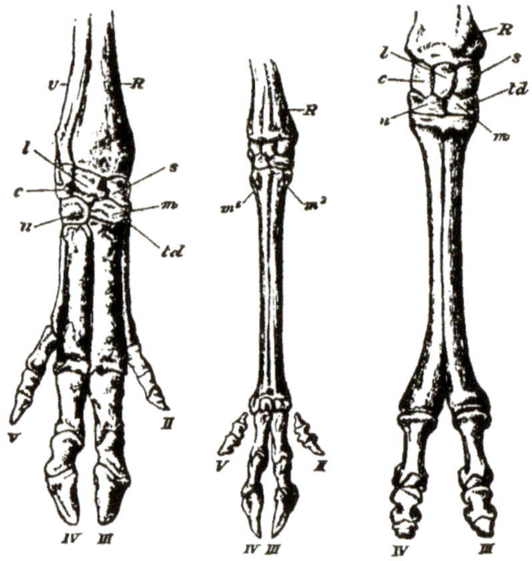

Bones of right fore feet of existing Artiodactyla. From left to right: Pig (Sus scrofa), Red deer (Cervus elaphus), and Camel (Camelus bactrianus). U = ulna, R = radius, c = cuneiform, l = lunar, s = Scaphoid, u = Unciform, m = Magnum, td = Trapezoid. In the Sheep and the Camel, the long compound bone, supporting the two main (or only) toes is the cannon bone.

A symbiotic relationship with microorganisms and a long digestive tract, with multiple stomach chambers, allows most even-toed ungulates to have an exclusively herbivorous diet, digesting plant matter (such as cellulose) that otherwise would provide little nutrition. The microorganisms also provide protein for the ungulates, while in turn the microbes receive a habitat and continuous supply of plant matter to digest.

The ungulates, or hoofed mammals, are divided into two living orders, Artiodactyla, the even-toed ungulates, and Perissodactyla, the odd-toed ungulates. As evident by the names of the orders—Artiodactyla comes from the Greek artios, meaning "even numbered" or "entire" and dactylos, meaning finger or toe, and Perrissodactyla from the Greek perissos, meaning "odd" and dactylos—the two groups are primarily distinguished by the feet.

All artiodactyls have an even number of functional toes, either two or four, on each foot (although there is conflicting information regarding the number of toes in the hind feet of a few species of peccaries in the Tayassuidae family). The symmetry of the foot passes between the middle two digits. In those artiodactyls with two main toes, the weight is borne on these two central elements, while the other toes are either reduced, vestigial, or absent, and when present are called dew claws or lateral hooves. In those artiodactyls with four weight-bearing (functional) toes, the toes form a spreading foot, with the nails at the end of the toes often enlarged. This is in contrast with the perissodactyls where the weight is born mostly or entirely by the middle toe.

Another key distinguishing feature is the shape of the astragalus. The astragalus is a tarsal (ankle) bone in the hind limb. It has deeply arched grooves where it articulates with the limb bones on either side (grooved joints). These grooves help in resisting lateral motion and also creates a double-pulley system that gives the foot greater flexibility and additionally increases the springiness of the lower hind limb. The double-pulley astragalus is a key feature in identifying artiodactyls in the fossil record.

While artiodactyls are the most abundant large land mammals today, there also are some small members. The mouse deer (Tragulidae family) of Southeast Asia stand no more than 35 centimeters (14 inches) at the shoulder and weigh less than one kilogram (two pounds). On the other hand, the largest member of the order, the common hippopotamus (Hippopotmus amphibius) reaches 4500 kilograms (10,000 pounds).

Artiodactyls vary greatly in form, with some with very long necks and other short; some with long faces and other short; some with a straight back, others with the front shoulders raised higher than the rump, and others with the rump higher than the shoulders; a tail that may be very short to long; and legs that may be long or short and slender to quite stout. Female artiodactyls have two to four teats, but members of the Suidae family have six to twelve teats.

Almost all species have some sort of weapons, whether unbranched horns, forked horns, antlers, or well-developed canines or tusks. These typically are largest in males and smaller or absent in females. The pelage consists of longer, stout guard hairs and shorter, normally finer underfur, although domestic sheep commonly have been bread so they lack the guard hairs, with the underfur providing wool.

Suborders

Artiodactyls traditionally have been placed into the three suborders of Suiformes, Tylopoda, and Ruminantia. The Suiformes classically have contained the three extant families of Suidae (pigs), Tayassuidae (peccaries and javelinas) and Hippopotamindae (hippopotamuses), although the hippos in many current taxonomies are placed in their own suborder of Cetancodonta. The only extant group in the Tylopoda is the family Camelidae (camels, llamas, and other camelids). The suborder Ruminantia contains the true ruminants, which are the extant families Giraffidae (giraffe and okapi), Cervidae (deer), Tragulidae (mouse deer and chevrotains), Antilocapridae (pronghorn), and Bovidae (antelopes, cattle, sheep, goats).

The group's suborders differ in various characteristics. Suina (pigs and peccaries) have retained four toes of fairly equal size, have simpler molars, short legs, and often have enlarged canine teeth that form tusks. Camelids and Ruminantia tend to be longer-legged, to walk on only the central two toes (though the outer two may survive as rarely-used dew-claws) and to have more complex cheek teeth well-suited to grinding up tough grasses.

Digestive System

Artiodactyls all have one or more digestive chambers located before the true stomach (abomasum).

Most members of the suborder Ruminantia have a four-chambered stomach, with the four chambers being the rumen, reticulum, omasum, and abomasum. This suborder includes such ruminating mammals as cattle, goats, sheep, giraffes, American Bison, European bison, yaks, water buffalo, deer, wildebeest, antelope, pronghorn, and nilgai.

However, the chevrotains (family Tragulidae) within Ruminantia have a three-chambered stomach. Likewise, members of the suborder Tylopoda (camels, alpacas, llamas) have a three-chambered stomach.

All of these animals are still considered "ruminants," even though the camelids are not part of Ruminantia. This is because the term ruminant simply means any even-toed ungulate that digests its food in two steps, first by softening it within the animal's first stomach, known as the rumen, then regurgitating the semi-digested mass, now known as cud, and chewing it again. Therefore, the term "ruminant" is not synonymous with Ruminantia. An alternative name for ruminants is "foregut fermentors," because bacterial fermentation takes place in the rumen, which is the first and largest chamber.

The pigs and peccaries have only one small chamber before the abomasum, and hippos (like the camelids and tragulids) have two chambers before the true stomach.

While hippopotamuses have a three-chambered stomach, they do not ruminate. They consume grass during the night and may cover large distances (up to 20 miles) to feed. They eat around 68 kg of food each night, also relying on microbes to break down plant material with cellulase.

Most species within Suina have a simple two-chambered stomach that allows an omnivorous diet; the babirusa, however, is a herbivore. They have extra maxillary teeth to allow proper mastication of plant material. Most of the fermentation occurs in the caecum with the help of cellulolytic microorganisms. Peccaries however have a complex stomach that contains four compartments. Microbial fermentation with the formation of high volatile fatty acid levels has been observed in the fore stomach. It has been proposed that their complex fore stomach is a means to slow digestive passage and increase digestive efficiency.

In the ruminants (Ruminantia and Tylopoda), food is regurgitated and rechewed then broken down by microbes in the stomach. After ingestion of plant material it is mixed with saliva in the rumen and reticulum and separates into layers of solid and liquid material. The solids lump together to form a bolus (also known as the cud), this is regurgitated by reticular contractions while the glottis is closed. When the bolus enters the mouth, the fluid is squeezed out with the tongue and reswallowed. The bolus is chewed slowly to completely mix it with saliva and to break down the particle size. Ingested food passes to the 'fermentation chamber' (rumen and reticulum) where it is kept in continual motion by rhythmic contractions of this organ. Cellulytic microbes (bacteria, protozoa, and fungi) produce cellulase, which is needed to break down the cellulose found in plant material. Without this mutual symbiosis ruminants would find plant material indigestible.

Not only are ruminants dependent on microorganisms for digesting cellulose, but the microorganisms also are the main protein source, with the ruminants digesting the microbes.

Habitat and Distribution

Even-toed ungulates are found on every continent but Antarctica; they were introduced to Australia and New Zealand by humans. While not native to the oceanic islands, domestic species have been introduced to many small remote islands.

Artiodactyls inhabit most ecosystems from arctic tundra to tropical forest, including deserts, valley floors, and mountaintops.

Behavior, Feeding and Reproduction

Artiodactyls are generally gregarious animals, living in groups, even if the groups are limited to two or three members. However, the sexes tend to remain separate, with adult males living apart from females and young.

Most artiodactyls are obligate herbivores that rely on plant matter for their diet. However, members of the Suidae and Tayassuidae are omnivores. And there are records of other artiodactyls occasionally eating the eggs and young of ground-nesting birds and other animal protein. The ancestors of the even-toed ungulates were believed to be omnivores that preferred plant material.

Members of the Suidae give birth to between four and eight young, with domestic pigs giving birth to more than 12 piglets per litter. However, other artiodactyls generally give birth to one, sometimes two young a year. All artiodactyl young are precocial, able to walk within hours after birth. One male mating with several females is the most common mating system, but temporary mating pairs may form and in a few species these mated pairs can stay together much of the year.

Relationship with Humans

The even-toed ungulates are of great historical and present economic and cultural value. They served as large mammal prey for early hunters. Cro-Magnons relied heavily on reindeer for food, skins, tools and weapons; with dropping temperatures and increased reindeer numbers at the end of the Pleistocene, they became the prey of choice. By around 12,500 years ago, reindeer remains accounted for 94 percent of bones and teeth found in a cave above the Céou River.

Today, many species of even-toed ungulates are still hunted for food and sport (deer, antelopes, African buffalo, wild sheep, etc.). In addition, the most important domestic livestock are artiodactyls, including cattle, goats, sheep, pigs, and camels. Sheep and goats likely were the first animals to be domesticated after the dog, perhaps 8,000 to 9,000 years ago. Cattletoday are the basis of a multi-billion dollar industry worldwide. The international trade in beef for 2000 was over $30 billion and represented only 23 percent of world beef production. Artiodactyls, both wild and domesticated, are used to provide meat, fur, milk, fertilizer, medicinal products, bones, antlers, horns, and serve as beasts of burden.

As with many mammal groups, even-toed ungulates first appeared during the Early Eocene (about 54 million years ago). In form they were rather like today's chevrotains: small, short-legged creatures that ate leaves and the soft parts of plants. By the Late Eocene (46 million years ago), the three modern suborders had already developed: Suina (the pig group); Tylopoda (the camel group); and Ruminantia (the goat and cattle group). Nevertheless, artiodactyls were far from dominant at that time: the odd-toed ungulates(ancestors of today's horses and rhinos) were much more successful and far more numerous. Even-toed ungulates survived in niche roles, usually occupying marginal habitats, and it is presumably at that time that they developed their complex digestive systems, which allowed them to survive on lower-grade food.

The appearance of grasses during the Eocene and their subsequent spread during the Miocene (about 20 million years ago) saw a major change: grasses are very difficult to eat and the even-toed ungulates with their highly-developed stomachs were better able to adapt to this coarse, low-nutrition diet, and soon replaced the odd-toed ungulates as the dominant terrestrial herbivores. Now-extinct Artiodactyla that developed during the Miocene include the species Ampelomeryx, Tauromeryx, Triceromeryx, and others.

Cetaceans are conisdered to have arisen from artiodactyls, and an early whale from 47-million year old Eocene deposits was found to have a double-pulley astragalus. In some taxonomies, the cetaceans and even-toed ungulates have been placed together in Cetartiodactyla as sister groups, although DNA analysis has shown cetaceans evolved from within Artiodactyla.

The most recent theory into the origins of hippopotamidae suggests that hippos and whales shared a common semi-aquatic ancestor that branched off from other artiodactyls around 60 million years ago. This hypothesized ancestral group likely split into two branches around 54 million years ago. One branch would evolve into cetaceans, possibly beginning with the proto-whale Pakicetus from 52 million years ago with other early whale ancestors collectively known as Archaeoceti, which eventually underwent aquatic adaptation into the completely aquatic cetaceans.

Pig

A pig is the even-toed ungulate animal within family Suidae.

Pigs are mammals with stocky bodies, flat snouts, small eyes and large ears. They are highly intelligent, social animals, and are found all over the world.

Pigs are in the Suidae family, which includes eight genera and 16 species. Among those species are wild boars, warthogs and pygmy hogs and domestic pigs. Pig, hog and boar essentially describe the same animal, but there are some distinctions. A boar is an uncastrated male domestic pig, but it also means a wild pig of any gender. A hog often means a domestic pig that weighs more than 120 lbs. (54 kilograms). Pigs are also called swine. Pigs were among the first animals to be domesticated, about 6,000 years ago in China.

Size

Pigs usually weigh between 300 and 700 lbs. (140 and 300 kilograms). Domestic pigs are often bred to be heavier. A hog named Reggie set a weight record of 1,335 lbs. (605.5 kg) in the Iowa State Fair's "Biggest Boar" contest in 2012.

This wild boar is a real porker. Although they're the wild ancestor of domesticated pigs, these guys, Sus scrofa, are much grumpier and can become aggressive if cornered.

Wild pigs vary greatly in size and weight. The largest boar is the giant forest hog (Hylochoerus meinertzhageni). It grows up to 6.9 feet (2.1 meters) long. The heaviest boar is the Eurasian wild pig (Sus scrofa), which grows to 710 lbs (320 kg).

The smallest boar is the pygmy hog (Sus salvanius). It grows to a length between 1.8 and 2.4 feett (55 to 71 centimeters) and 9.8 inches (25 cm) tall from hoof to shoulder. The pygmy hog only weighs 14.5 to 21 lbs (6.6 to 9.7 kg).

Habitat

Boars, pigs and hogs live all over the world, except for Australia, Antarctica, northern Africa and far northern Eurasia. For example, red river hogs (Potamochoerus porcus), also called bush pigs, are found in Africa; babirusas (Babyrousa babyrussa), or pig deer, are found in Indonesia; and Visayan warty pigs (Sus cebifrons) come from the Philippines.

Wild pigs typically live in grasslands, wetlands, rain forests, savannas, scrublands and temperate forests.

Habits

Pigs are very intelligent animals. Pigs are smarter than 3-year-old children. Domesticated pigs can be trained to do tricks and even use a litter box like a cat.

They are also very social. In the wild, pigs live in close-knit groups called sounders. Sounders consist of one male, many females and their young. Males without a sounder will form their own herds or be solitary. Bearded pigs can have sounders with up to 300 members.

Pigs communicate with a variety of grunts and squeaks. A short grunt, a longer growl and a loud roar will warn other pigs of approaching danger. The pigs' primary defense is speed, but when cornered, their tusks can be formidable weapons. Their lower tusks can get to be about 3 inches long (7 centimeters) and are razor sharp.

Diet

Pigs, boars and hogs are omnivores and will eat just about anything. Wild boars eat roots, fruit, rodents and small reptiles. Domestic pigs and hogs are fed feed that is made from corn, wheat, soy or barley. On small farms, pigs are often fed "slop," which consists of vegetable peels, fruit rinds and other leftover food items. Sometimes, free-range pigs find their own food.

Offspring

Female pigs, called cows or sows, give birth to offspring twice a year to a litter of around 12 young. Baby pigs are called piglets. At birth, piglets weigh around 2.5 lbs. (1.1 kg). Within a week, most piglets will double their weight. When they are two to four weeks old, the piglets are weaned.

Wild pigs can give birth to six for 14 piglets at a time. These piglets will stay in a nest for their first 10 days and are weaned after three months. Wild pigs live five to 20 years.

Classification/Taxonomy

Here is the taxonomy of pigs according to the Integrated Taxonomic Information System.

- Kingdom: Animalia
- Subkingdom: Bilateria
- Infrakingdom: Deuterostomia
- Phylum: Chordata
- Subphylum: Vertebrata
- Infraphylum: Gnathostomata
- Superclass: Tetrapoda
- Class: Mammalia
- Subclass: Theria
- Infraclass: Eutheria
- Order: Artiodactyla
- Family: Suidae

- Subfamily: Suinae
- Tribes: Babyrousini, Phacochoerini, Potamochoerini, Suini
- Genera: Babyrousa, Phacochoerus, Hylochoerus, Potamochoerus, Sus
- Species (in Sus genus):
  - Sus ahoenobarbus (Palawan pig)
  - Sus barbatus (bearded pig), with two subspecies
  - Sus bucculentus (Indochinese warty pig)
  - Sus cebifrons (Visayan warty pig), with two subspecies
  - Sus celebensis (Sulawesi warty pig), with three subspecies
  - Sus oliveri (Oliver's warty pig)
  - Sus philippensis (Philippine warty pig), with two subspecies
  - Sus salvanius (pygmy hog)
  - Sus scrofa (wild boar, pig, feral pig), with 16 subspecies, including Sus scrofa domesticus (domestic pig)
  - Sus verrucosus (Javan warty pig), with two subspecies

Conservation Status

Wild boars are not endangered, according to the International Union for Conservation of Nature's Red List of Threatened Species. They are listed as "least concern" due to the wild pig's "wide range, abundance, tolerance to habitat disturbance and presence in many protected areas."

Sulawesi warty pigs are listed as "near threatened"; bearded pigs, Palawan bearded pigs and Philippine warty pigs are "vulnerable"; Javan warty pigs are "endangered"; and Visayan warty pigs are "critically endangered." Hunting and habitat loss are cited as the causes of declining populations in these species.

Other Facts

- Domestic pigs have curly tails; wild pigs have straight tails.
- Eurasian wild pigs attack and eat snakes. The venom doesn't absorb as well because wild pigs have a layer of thick fat. The small amounts of venom that do make it through the fat are easily broken down.
- Pigs are very useful to the medical industry. Pig hearts are used as replacements for human hearts. Also, insulin and around 40 other medicines are made from pigs.
- Pigs are very clean and will keep their toilet area far away from where they lie down and eat.

- Pigs are very peaceful animals, rarely showing aggression, except when a mother with her young offspring is provoked or threatened.
- Pigs have a tremendous sense of smell. The large round disk of cartilage at the tip of the snout is connected to muscle that gives it extra flexibility and strength for rooting in the ground.
- Winston Churchill once said, "Dogs look up to man. Cats look down to man. Pigs look us straight in the eye and see an equal."
- The pig is the one of the 12 animals in the Chinese zodiac. The pig is seen to represent, fortune, honesty, happiness and virility.
- Bacon is one of the oldest processed meats in history. The Chinese began salting pork bellies as early as 1500 BC.
- The Spanish explorer Hernando DeSoto brought the first swine to the New World in 1539.
- A pig's squeal can range from 110 to 115 decibels. By comparison, noise from a motorcycle reaches about 100 decibels.
- A pig can run a seven-minute mile.
- The saying "living high on the hog" started among enlisted men in the U.S. Army, who received shoulder and leg cuts of pork while officers received the top loin cuts. So "living high on the hog" came to mean living well.
- "A pig in the poke" comes from a common trick in 17th-century England of trying to give away a cat to an unsuspecting "shopper" for a suckling pig (a young pig). When the buyer opened the poke (sack), he "let the cat out of the bag," and the trick was revealed.
- The phrase "pork barrel" politics" is derived from the pre-Civil War practice of distributing salt pork to the slaves from huge barrels. By the 1870s, congressmen were referring to regularly dipping into the "pork barrel" to obtaining funds for popular projects in their home districts.

Hippopotamus

The common hippopotamus (*Hippopotamus amphibius*), or hippo, is a large, mostly herbivorous, semiaquatic mammal native to sub-Saharan Africa, and one of only two extant species in the family Hippopotamidae, the other being the pygmy hippopotamus (*Choeropsis liberiensis* or *Hexaprotodon liberiensis*). The name comes from the ancient Greek for "river horse" . After the elephant and rhinoceros, the common hippopotamus is the third-largest type of land mammal and the heaviest extant artiodactyl. Despite their physical resemblance to pigs and other terrestrial even-toed ungulates, the closest living relatives of the Hippopotamidae are cetaceans (whales, dolphins, porpoises, etc.) from which they diverged about 55 million years ago.

Common hippos are recognisable by their barrel-shaped torsos, wide-opening mouths revealing large canine tusks, nearly hairless bodies, columnar legs and large size; adults average 1,500 kg

(3,310 lb) and 1,300 kg (2,870 lb) for males and females respectively. Despite its stocky shape and short legs, it is capable of running 30 km/h (19 mph) over short distances.

The common hippopotamus inhabits rivers, lakes and mangrove swamps, where territorial bulls preside over a stretch of river and groups of five to thirty females and young. During the day, they remain cool by staying in the water or mud; reproduction and childbirth both occur in water. They emerge at dusk to graze on grasses. While hippopotamuses rest near each other in the water, grazing is a solitary activity and hippos are not territorial on land. The hippopotamus is among the most dangerous animals in the world as it is highly aggressive and unpredictable. They are threatened by habitat loss and poaching for their meat and ivory canine teeth.

Description

Hippo's skull, showing the large canines and incisors used for fighting.

Hippopotami are among the largest living land mammals, being only smaller than elephants and some rhinoceroses. Mean adult weight is around 1,500 kg (3,310 lb) and 1,300 kg (2,870 lb) for males and females respectively, very large males can reach 2,000 kg (4,410 lb) and exceptional males weighing 2,660 kg (5,860 lb) and 3,200 kg (7,050 lb) have been reported. Male hippos appear to continue growing throughout their lives while females reach maximum weight at around age 25.

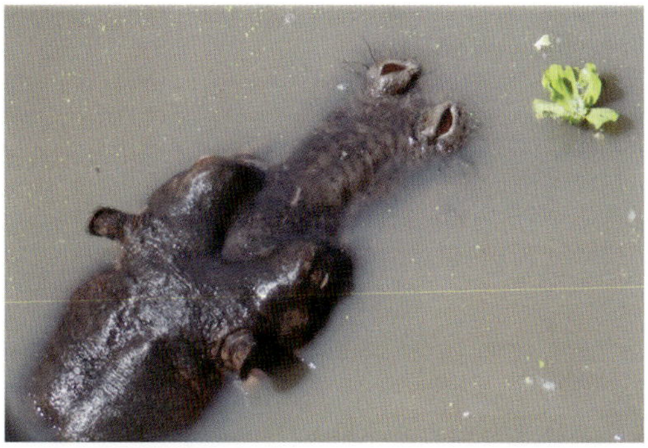

Mostly submerged hippo with exposed eyes, ears, and nostrils

Hippopotami have barrel-shaped bodies with short legs and long muzzles. Their skeletal structures are graviportal, adapted to carrying their enormous weight, and their specific gravity allows them to sink and move along the bottom of a river. Hippopotamuses have small legs (relative to other megafauna) because the water in which they live reduces the weight burden. Though they are bulky animals, hippopotamuses can gallop at 30 km/h (19 mph) on land but normally trot. They are incapable of jumping but do climb up steep banks. Despite being semiaquatic and having webbed feet, an adult hippo is not a particularly good swimmer nor can it float. It is rarely found in deep water; when it is, the animal moves by porpoise-like leaps from the bottom. The eyes, ears, and nostrils of hippos are placed high on the roof of their skulls. This allows these organs to remain above the surface while the rest of the body submerges. The testes of the males descend only partially and a scrotum is not present. In addition, the penis retracts into the body when not erect. The genitals of the female are unusual in that the vagina is ridged and two large diverticula protrude from the vulval vestibule. The function of these is unknown.

Characteristic "yawn" of a hippo.

The hippo's jaw is powered by a large masseter and a well-developed digastric; the latter loops up behind the former to the hyoid. The jaw hinge is located far back enough to allow the animal to open its mouth at almost 180°. A moderate folding of the orbicularis oris muscle allows the hippo to achieve such a gape without tearing any tissue. The bite force of an adult female has been measured as 8,100 newtons (1,800 lbf). Hippopotamus teeth sharpen themselves as they grind together. The lower canines and lower incisors are enlarged, especially in males, and grow continuously. The incisors can reach 40 cm (1 ft 4 in), while the canines reach up to 50 cm (1 ft 8 in). The canines and incisors are used for combat and play no role in feeding. Hippos rely on their broad horny lips to grasp and pull grasses which are then ground by the molars. The hippo is considered to be a pseudoruminant; it has a complex three-chambered stomach but does not "chew cud".

Completely submerged hippo

Unlike most other semiaquatic animals, the hippopotamus has very little hair. The skin is 6 cm (2 in) thick, providing it great protection against conspecifics and predators. By contrast, its subcutaneous fat layer is thin. The animals' upper parts are purplish-gray to blue-black, while the under parts and areas around the eyes and ears can be brownish-pink. Their skin secretes a natural sunscreen substance which is red-coloured. The secretion is sometimes referred to as "blood sweat", but is neither blood nor sweat. This secretion is initially colourless and turns red-orange within minutes, eventually becoming brown. Two distinct pigments have been identified in the secretions, one red (hipposudoric acid) and one orange (norhipposudoric acid). The two pigments are highly acidic compounds. They inhibit the growth of disease-causing bacteria, and their light absorption peaks in the ultraviolet range, creating a sunscreen effect. All hippos, even those with different diets, secrete the pigments, so it does not appear that food is the source of the pigments. Instead, the animals may synthesize the pigments from precursors such as the amino acid tyrosine. Nevertheless, this natural sunscreen cannot prevent the animal's skin from cracking if it stays out of water too long.

A hippo's lifespan is typically 40–50 years. Donna the Hippo was the oldest living hippo in captivity. She lived at the Mesker Park Zoo in Evansville, Indiana in the US until her death in 2012 at the age of 61.

Distribution and Status

Hippopotamus amphibius was widespread in North Africa and Europe during the Eemian and late Pleistocene until about 30,000 years ago. Archaeological evidence exists of its presence in the Levant, dating to less than 3,000 years ago. The species was common in Egypt's Nile region during antiquity, but has since been extirpated. Pliny the Elder writes that, in his time, the best location in Egypt for capturing this animal was in the Saite nome; the animal could still be found along the Damietta branch after the Arab Conquest in 639. Reports of the slaughter of the last hippopotamus in Natal Province were made at the end of the 19th century. Hippos are still found in the rivers and lakes of the northern Democratic Republic of the Congo, Uganda, Tanzania and Kenya, north through to Ethiopia, Somalia and Sudan, west to The Gambia, and south to South Africa.

Ugandan tribespeople with hippo slain for food

Incised hippopotamus ivory tusk, upper canine. Four holes around top

Genetic evidence suggests that common hippos in Africa experienced a marked population expansion during or after the Pleistocene, attributed to an increase in water bodies at the end of the era. These findings have important conservation implications as hippo populations across the continent are currently threatened by loss of access to fresh water. Hippos are also subject to unregulated hunting and poaching. In May 2006, the hippopotamus was identified as a vulnerable species on the IUCN Red List drawn up by the International Union for Conservation of Nature (IUCN), with an estimated population of between 125,000 and 150,000 hippos, a decline of between 7% and 20% since the IUCN's 1996 study. Zambia (40,000) and Tanzania (20,000–30,000) possess the largest populations.

The hippo population declined most dramatically in the Democratic Republic of the Congo. By 2005, the population in Virunga National Park had dropped to 800 or 900 from around 29,000 in the mid-1970s. The decline is attributed to the disruptions caused by the Second Congo War. The poachers are believed to be Mai-Mai rebels, poorly paid Congolese soldiers, and local militia groups. Reasons for poaching include the belief that hippos are harmful to society, as well as financial gain. However, as of 2016, the Virunga hippo population appears to have increased, possibly due to greater enforcement and cooperation between fishermen and park authorities. The sale of hippo meat is illegal, but black-market sales are difficult for Virunga National Park officers to track. Hippo meat is considered a delicacy in some areas of central Africa and the teeth have become a valued substitute for elephant ivory.

Invasive Potential

In the late 1980s, Colombian drug lord Pablo Escobar kept four hippos in a private menagerie at his residence in Hacienda Nápoles, 100 kiometres (62 mi) east of Medellín, Colombia, after buying them in New Orleans. They were deemed too difficult to seize and move after Escobar's death, and hence left on the untended estate. By 2007, the animals had multiplied to 16 and had taken to roaming the area for food in the nearby Magdalena River.

In 2009, two adults and one calf escaped the herd and, after attacking humans and killing cattle, one of the adults (called "Pepe") was killed by hunters under authorization of the local authorities. As of early 2014, 40 hippos have been reported to exist in Puerto Triunfo, Antioquia from the original four belonging to Escobar. Without management the population size is likely to more than double in next decade. The National Geographic Channel produced a documentary about them titled *Cocaine Hippos*. A report published in a Yale student magazine noted that local environmentalists are campaigning to protect the animals, although there is no clear plan for what will happen to them.

Behavior and Life History

Different from all other large land mammals, hippos are of semiaquatic habits, spending the day in lakes and rivers. They can be found in both savannah and forest areas. Proper habitat requires enough water to submerge in and grass nearby. Larger densities of the animals inhabit quiet waters with mostly firm, smooth sloping beaches. Males may be found in very small numbers in rapid waters in rocky gorges. Hippo mostly live in freshwater habitats, however populations in West Africa mostly inhabit estuarine waters and may even be found at sea. With the exception of eating, most of hippopotamuses' lives – from fighting with other hippos, mating, to parturition – occurs in the water. Hippos leave the water at dusk and travel inland, sometimes up to 10 km (6 mi), to graze on short grasses, their main source of food. They spend four to five hours grazing and can consume 68 kg (150 lb) of grass each night.

Hippopotamus out of the water after sunrise

Hippopotamus in the water

Hippopotamus feeding on grass

Like almost any herbivore, they consume other plants if presented with them, but their diet in nature consists almost entirely of grass, with only minimal consumption of aquatic plants. Hippos are born with sterile intestines, and require bacteria obtained from their mothers' feces to digest vegetation. Hippos have (albeit rarely) been filmed eating carrion, usually close to the water. There are other reports of meat-eating, and even cannibalism and predation. The stomach anatomy of a hippo is not suited to carnivory, and meat-eating is likely caused by aberrant behaviour or nutritional stress.

Hippos in the wild

Hippo defecation creates allochthonous deposits of organic matter along the river beds. These deposits have an unclear ecological function. A 2015 study concluded that hippo dung provides nutrients from terrestrial material for fish and aquatic invertebrates, while a 2018 study found that their dung can be toxic to aquatic life in large quantities, due to absorption of dissolved oxygen in water bodies. Because of their size and their habit of taking the same paths to feed, hippos can have a significant impact on the land across which they walk, both by keeping the land clear of vegetation and depressing the ground. Over prolonged periods, hippos can divert the paths of swamps and channels.

A hippopotamus walking on the grass land in Seregeti National Park in the morning

Adult hippos move at speeds up to 8 km/h (5 mph) in water; typically resurfacing to breathe every three to five minutes. The young have to breathe every two to three minutes. The process of surfacing and breathing is subconscious. A hippo sleeping underwater rises and breathes without waking up. A hippo closes its nostrils when it submerges into the water. As with fish and turtles on a coral reef, hippos occasionally visit cleaning stations and signal, by opening their mouths wide, their readiness for being cleaned of parasites by certain species of fishes. This is an example of mutualism in which the hippo benefits from the cleaning, while the fish receive food.

Hippopotamus coexist with a variety of formidable predators. Nile crocodiles, lions and spotted hyenas are known to prey on young hippos. However, due to their aggression and size, adult hippopotamus are not usually preyed upon by other animals. Cases where large lion prides have successfully preyed on adult hippopotamus have been reported; however, this predation is generally rare. Lions occasionally prey on adults at Gorongosa National Park and calves are taken at Virunga. Crocodiles are frequent targets of hippo aggression, probably because they often inhabit the same riparian habitats; crocodiles may be either aggressively displaced or killed by hippopotamuses.

In turn, beyond cases of killing the seldom unguarded hippo calf, very large Nile crocodiles have been verified to occasionally prey on "half-grown" hippopotamuses and anecdotally perhaps adult female hippos. Also, aggregations of crocodiles that have been seen to dispatch still living bull hippopotamus that have been previously injured in mating battles with other bulls.

Social Spacing

Hippopotamus pod

Studying the interaction of male and female hippopotami has long been complicated because hippos are not sexually dimorphic; thus females and young males are almost indistinguishable in the field. Although hippos lie close to each other, they do not seem to form social bonds except between mothers and daughters, and they are not social animals. The reason they huddle close together is unknown. Hippopotamuses are territorial only in water, where a bull presides over a small stretch of river, on average 250 m (270 yd) in length, and containing 10 females. The largest pods can contain over 100 hippos. Younger bachelors are allowed in a bull's stretch, as long as they behave submissively toward the bull. The territories of hippos exist to establish mating rights. Within the pods, the hippos tend to segregate by gender. Bachelors lounge near other bachelors, females with other females, and the bull on his own. When hippos emerge from the water to graze, they do so individually.

Males fighting

Hippos mark their territory by defecation. While depositing the faeces, hippos spin their tails to distribute their excrement over a greater area. "Yawning" serves as a threat display. When fighting, male hippos use their incisors to block each other's attacks and their large canines to inflict injuries. When hippos become over-populated or a habitat is reduced, bulls sometimes attempt infanticide, but this behaviour is not common under normal conditions. Incidents of hippo cannibalism have been documented, but this is believed to be the behaviour of distressed or sick hippos.

Hippopotami appear to communicate vocally, through grunts and bellows, and they may practice echolocation, but the purpose of these vocalizations is currently unknown. Hippos have the unique ability to hold their heads partially above the water and send out a cry that travels through both water and air; individuals respond above and under water. Hippos will also express threat and alarm with exhalations.

Reproduction

A juvenile at the Zoo Basel

Mother with young

Female hippos reach sexual maturity at five to six years and have a gestation period of eight months. A study of endocrine systems revealed that female hippopotamuses may begin puberty as early as three or four years. Males reach maturity at around 7.5 yr. A study of hippopotami reproductive behaviour in Uganda showed that peak conceptions occurred during the end of the wet season in the summer, and peak births occurred toward the beginning of the wet season in late winter. This is because of the female's oestrous cycle; as with most large mammals, male hippopotamus spermatozoa is active year-round. Studies of hippos in Zambia and South Africa also showed evidence of births occurring at the start of the wet season. After becoming pregnant, a female hippopotamus will typically not begin ovulation again for 17 months.

Preserved hippo fetus

Mating occurs in the water, with the female submerged for most of the encounter, her head emerging periodically to draw breath. Female hippos isolate themselves to give birth and return within 10–14 days. Baby hippos are born underwater at a weight between 25 and 50 kg (55 and 110 lb) and an average length of around 127 cm (4.17 ft), and must swim to the surface to take their first breaths. A mother typically gives birth to only one calf, although twins also occur. The young often rest on their mothers' backs when the water is too deep for them, and they swim under water to suckle. They suckle on land when the mother leaves the water.

Mother hippos are very protective of their young and may keep others at a distance. However, calves are occasionally left in nurseries which are guarded by one or a few adults. Calves in nurseries engage in playfights. Weaning starts between six and eight months after birth, and most calves are fully weaned after a year. Like many other large mammals, hippos are described as K-strategists, in this case typically producing just one large, well-developed infant every couple of years (rather than many small, poorly developed young several times per year as is common among small mammals such as rodents).

Hippos and Humans

The Hippopotamus and Crocodile Hunt

The earliest evidence of human interaction with hippos comes from butchery cut marks on hippo bones at Bouri Formation dated around 160,000 years ago. Later rock paintings and engravings showing hippos being hunted have been found in the mountains of the central Sahara dated 4,000–5,000 years ago near Djanet in the Tassili n'Ajjer Mountains. The ancient Egyptians recognised the hippo as a ferocious denizen of the Nile and representations on the tombs of nobles show that the animals were hunted.

The hippopotamus was also known to the Greeks and Romans. The Greek historian Herodotus described the hippopotamus in *The Histories* and the Roman naturalist Pliny the Elder wrote about the hippopotamus in his encyclopedia *Naturalis Historia*. The Yoruba people called the hippopotamus *erinmi* which means "elephant of the water". Zulu warriors preferred to be as brave as a hippopotamus, since even lions were not considered to match its courage.

Hippopotamus, Egyptian, Middle Kingdom

In the U.S., Representative Robert F. Broussard of Louisiana introduced the "American Hippo bill" in 1910 to authorize the importation and release of hippopotamus into the bayous of Louisiana. Broussard argued that the hippopotamus would eat the invasive water hyacinth that was clogging the rivers and also produce meat to help solve the American meat crisis. The chief collaborators and proponents of Broussard's bill were Major Frederick Russell Burnham and Captain Fritz Duquesne. Former President Theodore Roosevelt backed the plan, as did the U.S. Department of Agriculture, *The Washington Post*, and *The New York Times* which praised the taste of hippopotamus as "lake cow bacon". The "American Hippo Bill" fell just short of being passed.

Attacks on Humans

The hippopotamus is considered to be very aggressive and has frequently been reported as charging and attacking boats. Small boats can be capsized by hippos and passengers can be injured or killed by the animals or drown. In one case in Niger, a boat was capsized by a hippo and 13 people were killed. As hippopotamuses will often engage in raiding nearby crops if the opportunity arises, humans may also come in conflict with them on these occasions, with potential for fatalities on both sides.

Hippos in Zoos

Hippopotamuses have long been popular zoo animals. The first zoo hippo in modern history was Obaysch, who arrived at the London Zoo on 25 May 1850, where he attracted up to 10,000 visitors a day and inspired a popular song, the "Hippopotamus Polka". Hippos generally breed well in captivity; birth rates are lower than in the wild, but this is attributed to zoos wanting to limit births, since hippos are relatively expensive to maintain.

Obaysch lounging at the London Zoo

A hippopotamus at Woodland Park Zoo in Seattle, Washington

Like many zoo animals, hippos were traditionally displayed in concrete exhibits. In the case of hippos, they usually had a pool of water and patch of grass. In the 1980s, zoo exhibits increasingly reflected native habitats. The Toledo Zoo Hippoquarium, features a 360,000 gallon pool. In 1987, the Toledo Zoo saw the first underwater birth by a captive hippo. The exhibit was so popular, the hippos became the logo of the Toledo Zoo.

Giraffe

The giraffes are a even-toed ungulate mammals. Giraffes (Giraffa camelopardalis) are the world's tallest mammals. They are well known for their long necks, long legs, and spotted patterns.

Behavior and Ecology

Habitat and Feeding

A Masai giraffe extending its tongue to feed, in Tanzania. Its tongue, lips and palate are tough enough to deal with sharp thorns in trees.

Giraffes usually inhabit savannahs and open woodlands. They prefer Acacieae, *Commiphora*, *Combretum* and open *Terminalia* woodlands over denser environments like *Brachystegia* woodlands. The Angolan giraffe can be found in desert environments. Giraffes browse on the twigs of trees, preferring trees of the subfamily Acacieae and the genera *Commiphora* and *Terminalia*, which are important sources of calcium and protein to sustain the giraffe's growth rate. They also feed on shrubs, grass and fruit. A giraffe eats around 34 kg (75 lb) of foliage daily. When stressed, giraffes may chew the bark off branches. Although herbivorous, the giraffe has been known to visit carcasses and lick dried meat off bones.

During the wet season, food is abundant and giraffes are more spread out, while during the dry season, they gather around the remaining evergreen trees and bushes. Mothers tend to feed in open areas, presumably to make it easier to detect predators, although this may reduce their feeding efficiency. As a ruminant, the giraffe first chews its food, then swallows it for processing and then visibly passes the half-digested cud up the neck and back into the mouth to chew again. It is common for a giraffe to salivate while feeding. The giraffe requires less food than many other herbivores because the foliage it eats has more concentrated nutrients and it has a more efficient digestive system. The animal's faeces come in the form of small pellets. When it has access to water, a giraffe drinks at intervals no longer than three days.

Giraffes have a great effect on the trees that they feed on, delaying the growth of young trees for some years and giving "waistlines" to trees that are too tall. Feeding is at its highest during the first and last hours of daytime. Between these hours, giraffes mostly stand and ruminate. Rumination is the dominant activity during the night, when it is mostly done lying down.

Social Life

Gathering of female southern giraffes in Tswalu Kalahari Reserve, South Africa. These animals commonly gather in herds.

Giraffes are usually found in groups. Traditionally, the composition of these groups has been described as open and ever-changing. Giraffes were thought to have few social bonds and for research purposes, a "group" has been defined as "a collection of individuals that are less than a kilometre apart and moving in the same general direction." More recent studies have found that giraffes do have long-term social associations and may form groups or pairs based on kinship, sex or other factors. These groups may regularly associate with one another in larger communities or sub-communities within a fission–fusion society. The number of giraffes in a group can range up to 44 individuals.

Giraffe groups tend to be sex-segregated although mixed-sex groups made of adult females and young males are known to occur. Particularly stable giraffe groups are those made of mothers and their young, which can last weeks or months. Social cohesion in these groups is maintained by the bonds formed between calves. Female association appears to be based on space-use and individuals may be matrilineally related. In general, females are more selective than males in who they associate with in regards to individuals of the same sex. Young males also form groups and will engage in playfights. However, as they get older males become more solitary but may also associate in pairs or with female groups. Giraffes are not territorial, but they have home ranges. Male giraffes occasionally wander far from areas that they normally frequent.

Although generally quiet and non-vocal, giraffes have been heard to communicate using various sounds. During courtship, males emit loud coughs. Females call their young by bellowing. Calves will emit snorts, bleats, mooing and mewing sounds. Giraffes also snore, hiss, moan, grunt and make flute-like sounds, and possibly communicate over long distances using infrasound—though this is disputed. During nighttime, giraffes appear to hum to each other above the infrasound range for purposes which are unclear.

Reproduction and Parental Care

Southern giraffes courting (above) and mating in Namibia. Generally, only dominant males are able to mate with females.

Reproduction in giraffes is broadly polygamous: a few older males mate with the fertile females. Male giraffes assess female fertility by tasting the female's urine to detect oestrus, in a multi-step process known as the flehmen response. Males prefer young adult females over juveniles and older adults. Once an oestrous female is detected, the male will attempt to court her. When courting, dominant males will keep subordinate ones at bay. A courting male may lick a female's tail, rest his head and neck on her body or nudge her with his horns. During copulation, the male stands on his hind legs with his head held up and his front legs resting on the female's sides.

Giraffe gestation lasts 400–460 days, after which a single calf is normally born, although twins occur on rare occasions. The mother gives birth standing up. The calf emerges head and front legs first, having broken through the fetal membranes, and falls to the ground, severing the umbilical cord. The mother then grooms the newborn and helps it stand up. A newborn giraffe is 1.7–2 m (5.6–6.6 ft) tall. Within a few hours of birth, the calf can run around and is almost indistinguishable from a one-week-old. However, for the first 1–3 weeks, it spends most of its time hiding; its coat pattern providing camouflage. The ossicones, which have lain flat while it was in the womb, become erect within a few days.

Mother Masai giraffe and calves feeding in Kenya. It is mostly the females that raise young, and they may gather in nursery herds.

Mothers with calves will gather in nursery herds, moving or browsing together. Mothers in such a group may sometimes leave their calves with one female while they forage and drink elsewhere. This is known as a "calving pool". Adult males play almost no role in raising the young, although they appear to have friendly interactions. Calves are at risk of predation, and a mother giraffe will stand over her calf and kick at an approaching predator. Females watching calving pools will only alert their own young if they detect a disturbance, although the others will take notice and follow.

The length time in which offspring stay with their mother varies, though it can last until the female's next calving. Likewise, calves may suckle for only a month or as long as a year. Females become sexually mature when they are four years old, while males become mature at four or five years. Spermatogenesis in male giraffes begins at three to four years of age. Males must wait until they are at least seven years old to gain the opportunity to mate.

Necking

Male giraffes use their necks as weapons in combat, a behaviour known as "necking". Necking is used to establish dominance and males that win necking bouts have greater reproductive success. This behaviour occurs at low or high intensity. In low intensity necking, the combatants rub and lean against each other. The male that can hold itself more erect wins the bout. In high intensity necking, the combatants will spread their front legs and swing their necks at each other, attempting to land blows with their ossicones. The contestants will try to dodge each other's blows and then get ready to counter. The power of a blow depends on the weight of the skull and the arc of the swing. A necking duel can last more than half an hour, depending on how well matched the combatants are. Although most fights do not lead to serious injury, there have been records of broken jaws, broken necks, and even deaths.

Here, male southern giraffes engage in low intensity necking to establish dominance, in Ithala Game Reserve, Kwa-Zulu-Natal, South Africa.

The more active kind of necking, engaged with here by southern giraffes in Namibia.

After a duel, it is common for two male giraffes to caress and court each other. Such interactions between males have been found to be more frequent than heterosexual coupling. In one study, up to 94 percent of observed mounting incidents took place between males. The proportion of same-sex activities varied from 30–75 percent. Only one percent of same-sex mounting incidents occurred between females.

Relationship with Humans

Humans have interacted with giraffes for millennia. The San people of southern Africa have medicine dances named after some animals; the giraffe dance is performed to treat head ailments. How the giraffe got its height has been the subject of various African folktales, including one from eastern Africa which explains that the giraffe grew tall from eating too many magic herbs. Giraffes were depicted in art throughout the African continent, including that of the Kiffians, Egyptians and Meroë Nubians. The Kiffians were responsible for a life-size rock engraving of two giraffes that has been called the "world's largest rock art petroglyph". The Egyptians gave the giraffe its own hieroglyph, named 'sr' in Old Egyptian and 'mmy' in later periods. They also kept giraffes as pets and shipped them around the Mediterranean.

San rock art in Namibia depicting a giraffe

Painting of a giraffe imported to China during the Ming dynasty

The giraffe was also known to the Greeks and Romans, who believed that it was an unnatural hybrid of a camel and a leopard and called it *camelopardalis*. The giraffe was among the many animals collected and displayed by the Romans. The first one in Rome was brought in by Julius Caesar in 46 BC and exhibited to the public. With the fall of the Western Roman Empire, the housing of giraffes in Europe declined. During the Middle Ages, giraffes were known to Europeans through contact with the Arabs, who revered the giraffe for its peculiar appearance.

Individual captive giraffes were given celebrity status throughout history. In 1414, a giraffe was shipped from Malindi to Bengal. It was then taken to China by explorer Zheng He and placed in a Ming dynasty zoo. The animal was a source of fascination for the Chinese people, who associated it with the mythical Qilin. The Medici giraffe was a giraffe presented to Lorenzo de' Medici in 1486. It caused a great stir on its arrival in Florence. Zarafa, another famous giraffe, was brought from Egypt to Paris in the early 19th century as a gift from Muhammad Ali of Egypt to Charles X of France. A sensation, the giraffe was the subject of numerous memorabilia or "giraffanalia".

Giraffes continue to have a presence in modern culture. Salvador Dalí depicted them with burning manes in some of his surrealist paintings. Dali considered the giraffe to be a symbol of masculinity, and a flaming giraffe was meant to be a "masculine cosmic apocalyptic monster". Several children's books feature the giraffe, including David A. Ufer's *The Giraffe Who Was Afraid of Heights*, Giles Andreae's *Giraffes Can't Dance* and Roald Dahl's *The Giraffe and the Pelly and Me*. Giraffes have appeared in animated films, as minor characters in Disney's *The Lion King* and *Dumbo*, and in more prominent roles in *The Wild* and in the *Madagascar* films. Sophie the Giraffe has been a popular teether since 1961. Another famous fictional giraffe is the Toys "R" Us mascot Geoffrey the Giraffe.

The giraffe has also been used for some scientific experiments and discoveries. Scientists have looked at the properties of giraffe skin when developing suits for astronauts and fighter pilots because the people in these professions are in danger of passing out if blood rushes to their legs. Computer scientists have modeled the coat patterns of several subspecies using reaction–diffusion mechanisms.

The constellation of Camelopardalis, introduced in the seventeenth century, depicts a giraffe. The Tswana people of Botswana traditionally see the constellation Crux as two giraffes – Acrux and Mimosa forming a male, and Gacrux and Delta Crucis forming the female.

Exploitation and Conservation Status

Masai giraffe killed by tribesmen in German East Africa during the early 20th century

Giraffes were probably common targets for hunters throughout Africa. Different parts of their bodies were used for different purposes. Their meat was used for food. The tail hairs served as flyswatters, bracelets, necklaces and thread. Shields, sandals and drums were made using the skin, and the strings of musical instruments were from the tendons. The smoke from burning giraffe skins was used by the medicine men of Buganda to treat nose bleeds. The Humr people of Sudan consume the drink Umm Nyolokh; which is created from the liver and marrow of giraffes. Umm

Nyolokh often contains DMT and other psychoactive substances from plants the giraffes eat such as Acacia; and is known to cause hallucinations of giraffes, believed to be the giraffes' ghosts by the Humr. In the 19th century, European explorers began to hunt them for sport. Habitat destruction has hurt the giraffe, too: in the Sahel, the need for firewood and grazing room for livestock has led to deforestation. Normally, giraffes can coexist with livestock, since they do not directly compete with them. In 2017, severe droughts in northern Kenya have led to increased tensions over land and the killing of wildlife by herders, with giraffe populations being particularly hit.

Aerial survey is the most common method of monitoring giraffe population trends in the vast roadless tracts of African landscapes, but aerial methods are known to undercount giraffes. Ground-based survey methods are more accurate and should be used in conjunction with aerial surveys to make accurate estimates of population sizes and trends. In 2010, giraffes were assessed as Least Concern from a conservation perspective by the International Union for Conservation of Nature (IUCN), but the 2016 assessment categorized giraffes as Vulnerable. Giraffes have been extirpated from much of their historic range including Eritrea, Guinea, Mauritania and Senegal. They may also have disappeared from Angola, Mali, and Nigeria, but have been introduced to Rwanda and Swaziland. Two subspecies, the West African giraffe and the Rothschild giraffe, have been classified as endangered, as wild populations of each of them number in the hundreds.

In 1997, Jonathan Kingdon suggested that the Nubian giraffe was the most threatened of all giraffes; as of 2010, it may number fewer than 250, although this estimate is uncertain. Private game reserves have contributed to the preservation of giraffe populations in southern Africa. Giraffe Manor is a popular hotel in Nairobi that also serves as sanctuary for Rothschild's giraffes. The giraffe is a protected species in most of its range. It is the national animal of Tanzania, and is protected by law. Unauthorised killing can result in imprisonment. The UN backed Convention of Migratory Species selected giraffes for protection in 2017. In 1999, it was estimated that over 140,000 giraffes existed in the wild, estimations as of 2016 indicate that there are approximately 97,500 members of *Giraffa* in the wild, down from 155,000 in 1985, with around 1,144 in captivity.

Camel

Camels are even-toed ungulate mammals with long legs, a big-lipped snout and a humped back. There are two types of camels: dromedary camels, which have one hump, and Bactrian camels, which have two humps. Camels' humps consist of stored fat, which they can metabolize when food and water is scarce.

In addition to their humps, camels have other ways to adapt to their environment. They have a third, clear eyelid that protects their eyes from blowing sand. Two rows of long lashes also protect their eyes. Sand up the nose can be a problem, but not for camels. They can shut their nostrils during sand storms.

Humans have used camels as a means of transport for thousands of years. They can carry about 375 to 600 lbs. (170 to 270 kilograms). This earned these beasts of burden a nickname, "ships of the desert." Domestic camels are often the main source of meat, milk and even leather or wool products.

A camel feeds her calf.

Size

Most camels tower above humans. A Bactrian camel, grows to a shoulder height of 6 feet (1.8 meters) and a body length of 10 feet (3 m). They normally weigh 1,320 to 2,200 lbs. (600 to 1,000 kg) when they are fully grown.

Dromedary camels get up to about 6.5 feet (2 m) tall at the shoulder and weigh 880 to 1,325 lbs. (400 to 600 kg).

Diet

Camels aren't picky about what they eat. Their thick lips allow them to eat things that most other animals couldn't, such as thorny plants. Camels are herbivores, though, so you won't find them eating meat.

Filling up on water, when it's available, is very important for camels. They can drink 30 gallons (113 liters) of water in just 13 minutes. Their bodies rehydrate faster than any other mammal.

When there is little food and water, the camel's hump fat releases water; 9.3 grams of fat releases 1.13 grams of water. Camels can survive up to six months without food or water.

Habitat

The two types of camel are found in different parts of the world. The dromedary camel, also called an Arabian camel, can be found in North Africa and the Middle East. The Bactrian camel lives in Central Asia. No matter the type, camels are usually found in the desert, prairie or steppe. Though many people think that camels only live in hot climates, they do well in temperature ranges from 20 degrees F (minus 29 degrees C) to 120 degrees F (49 degrees C).

Habits

Camels like to stay together in groups called herds. The herds are led by a dominant male, while many of the other males form their own herd called a bachelor herd. Camels are very social and like to greet each other by blowing in each other's faces.

Offspring

After a gestation of 12 to 14 months, a mother camel will find a private spot to have her young. Female camels usually only have one baby, but sometimes camels have twins.

Baby camels are called calves. The newborn calf is able to walk within 30 minutes, though the two won't rejoin the herd until around two weeks later. Camels become fully mature when they are 7 years old. Camels live around 17 years.

Classification/Taxonomy

Dromedary camels (Camelus dromedarius) and the domestic Bactrian camel (Camelus bactrianus) were named in 1758 by Swedish zoologist Carl Linnaeus, who only knew of the domestic variety. Wild Bactrian camels (Camelus ferus) were discovered in 1878 by Nikolai Prejevalsky, a Russian geographer who explored Mongolia and Tibet.

For many years, the wild Bactrian was thought to be a subspecies of the domestic Bactrian. However, in recent years, DNA analysis confirmed that C. ferus was a separate species. The main difference between the two species is that the wild Bactrian has three more chromosome pairs than the domestic Bactrian.

Here is the classification of camels:

- Kingdom: Animalia
- Subkingdom: Bilateria
- Infrakingdom: Deuterostomia
- Phylum: Chordata
- Subphylum: Vertebrata
- Infraphylum: Gnathostomata
- Superclass: Tetrapoda
- Class: Mammalia
- Subclass: Theria
- Infraclass: Eutheria

- Order: Artiodactyla
- Family: Camelidae
- Genus: Camelus
- Species:
    - Camelus bactrianus (Bactrian camel)
    - Camelus dromedarius (one-humped camel)
- Subspecies:
    - Camelus bactrianus bactrianus
    - Camelus bactrianus ferus (wild Bactrian camel)

Conservation Status

Camelus dromedarius at the Singapore Zoo

The wild Bactrian camel is considered critically endangered by the International Union for Conservation of Nature and has a population that is decreasing. In fact, wild camels are one of the most endangered large mammals. There are fewer than 1,000 wild camels alive.

Other Facts

Camels can run at 25 mph (40 kph) for long periods. If their owner is in a hurry, they can kick their speed up to 40 mph (67 kph).

The camel's hump is like a storage container. When camels use their stored fat, their hump will diminish. When they eat and drink again the hump will refill with fat.

Camels have oval-shaped red blood cells that help continue blood flow during times when water is scarce.

Camels are known for spitting on people. In fact, the animals are throwing up the contents of their stomach along with spit. This is a defense tactic when the animals feel threatened.

The large beasts make a variety of moans, groans and deep, throaty bellows. One of the camel's noises was even used to voice the character Chewbacca in the Star Wars movies.

Moose

Moose, (Alces alces) is the largest, even-toed ungulate member of the deer family Cervidae (order Artiodactyla). Moose are striking in appearance because of their towering size, black colour, long legs, pendulous muzzle, and dangling hairy dewlap (called a bell) and the immense, wide, flat antlers of old bulls. The name moose is common in North America; it is derived from the word moosh ("stripper and eater of bark") in the Algonquian language of the Montagnais (Innu) Indians of Quebec, Canada. In Europe moose are called elk.

Moose inhabit the northern parts of North America and Eurasia. In North America four subspecies are recognized, including the eastern moose (A. alces americana), which inhabits eastern Canada and the northeastern United States; the northwestern moose (A. alces andersoni), which inhabits central Canada and North Dakota, Minnesota, and northern Michigan; the Shiras moose (A. alces shirasi), which inhabits the Rocky Mountains in the United States and Canada; and the Alaskan moose (A. alces gigas), which inhabits Alaska and northwestern Canada. Although not widely accepted, some classifications also recognize several Eurasian subspecies, including the European moose (A. alces alces); the Siberian, or Yakut, moose (A. alces pfizenmayeri); the west Siberian, or Ussuri, moose (A. alces cameloides); and the east Siberian, or Kolyma, moose (A. alces buturlini). In addition to differences in geographical distribution, the different subspecies of moose are further distinguished by features such as size, pelage, and antler characteristics. The differences in regional body sizes appears to reflect adaptation to local conditions. The largest moose specimens are found in Alaska and eastern Siberia; there bulls weigh 600 kg (1,300 pounds) and stand 2 metres (7 feet) tall at the shoulder. The smallest moose are found in its southernmost populations in Wyoming and Manchuria, where large bulls weigh 300–350 kg (660–770 pounds).

Moose (Alces alces) at the edge of a frozen river

Moose primarily exploit plant communities of deciduous shrubs that have been disturbed by flooding, avalanches, or forest fires. They are avid visitors to mineral licks. In winter they may also avidly consume conifers such as fir and yew. In areas of very deep snow, moose may tramp a system of trails called a "moose yard." In summer they may also consume large amounts of aquatic vegetation. The large, mobile, sensitive muzzle appears to be a specialized feeding organ that allows moose to exploit the large stocks of submerged aquatic vegetation in shallow lakes and streams. Moose may dive and stay up to 50 seconds underwater while feeding. Even calves are excellent swimmers.

Moose (Alces alces) with fully developed antlers

Moose are bold and readily defend themselves against large carnivores. During calving season, moose cows face grizzly and black bears. In late winter when the snow is deep and moose cannot flee, they defend themselves against wolf packs. They choose hard, level ground with little snow for maneuverability, such as ridges blown free of snow or frozen lakes with a thin cover of snow. When hindered by deep snow, they back into dense conifers to protect their vulnerable inguinal region and lower haunches from attacks by wolves. They may then charge the wolves and attack them by slapping them with their front legs and kicking them with their hind legs. These blows are powerful enough to kill wolves.

Moose in a forest in New Hampshire

Moose have killed humans. In Siberia, hunters armed with muzzle-loading guns feared wounded moose far more than they feared the large brown bear. Due to the thick skin on its head and neck and its dense skull, an attacking moose could not be readily stopped with a small, round rifle ball of soft lead.

Moose normally escape predators by trotting at high speed, which forces the pursuing smaller predators into expensive and tiring jumping but which costs a moose relatively little energy. It readily comes to bay but on its terms: it chooses low water where wolves are hampered in their movements. Although moose are excellent swimmers, it does not choose deeper water, because northern wolves have relatively large paws and so are also excellent swimmers. Predation by wolves and bears removes the infirm but may also severely deplete healthy calves, despite the spirited defense of their mothers.

Moose (Alces alces) Crossing a river

Moose mate in September so that the calves may be born in June to take advantage of spring vegetation. The antlers are shed of the blood-engorged skin called velvet in late August, and the bulls are in rut by the first week of September. Rutting bulls search widely for females, but the bulls may also attract females with the smell of their urine. They paw rutting pits with their forelegs, urinate into them, and splash the urine-soaked muck onto their hairy bells. Cows in turn may call to attract bulls. Actively rutting bulls appear to receive more than 50 punctures per mating season, but they are protected by a thick skin on the front and the neck. Rutting is expensive, as bulls lose virtually all of their body fat and their festering wounds must heal.

Because of their large body size, moose have a long gestation period of about 230 days. Twins are not uncommon. The young are born tan in color, which contrasts sharply with the dark color of adults. They grow very fast but still require maternal protection against wolves in winter. They are driven off by their mother shortly before she gives birth again. The dispersed yearlings roam in search of new living space.

Young calf moose in human hands tame readily and emerge as surprisingly intelligent, mischievous, but utterly loyal creatures. As mounts and beasts of burden, moose are superior to horses in muskeg and taiga. In the late 19th and early 20th centuries, moose became scarce owing to severe exploitation in unsettled times in Eurasia and to uncontrolled market hunting in North America. However, they responded readily to protection and management. Today moose are abundant in Eurasia and North America and are a cherished game animal. (The muzzle of the moose is considered a delicacy.) However, with the restoration of a predator fauna in North America, moose are again declining.

Mountain Goat

Despite its name, the mountain goat is actually a member of the antelope family. It has a long face, long black horns and a short tail. Both males and females have beard-like hair on their chins. The mountain goat sports a coat of wooly, white fur that keeps it warm at high elevations. This coat has a double layer for added warmth during winter—the overcoat molts, or falls off, during summer time.

Known for their agility, mountain goats are most often seen scaling steep, rocky ledges. This extreme alpine environment provides them with adequate protection from predators. Strong muscular forequarters and pliable hooves with soft rubbery pads help them maintain traction on craggy rock surfaces and survive in harsh conditions.

Alpine Adaptations

These surefooted beasts inhabit many of North America's most spectacular alpine environments. They often appear at precipitous heights, from Alaska to the U.S. Rocky Mountains, showcasing climbing abilities that leave other animals, including most humans, far below. Mountain goats have cloven hooves with two toes that spread wide to improve balance. Rough pads on the bottom of each toe provide the grip of a natural climbing shoe. Mountain goats are powerful but nimble and can jump nearly 12 feet in a single bound.

Mountain goats have distinctive beards and long, warm coats to protect them from cold temperatures and biting mountain winds. Their dazzling white coats provide good camouflage on the snowy heights. During the more moderate summer season goats shed this coat.

Reproduction

Female goats (called nannies) spend much of the year in herds with their young (called kids). These groups may include as many as 20 animals. Males (known as billies) usually live alone or with one or two other male goats. Both sexes boast beautiful pointed horns, and in mating season billies will sometimes use them to battle rivals for prospective mates.

In the spring, a nanny goat gives birth to one kid (sometimes two), which must be on its feet within minutes of arrival into its sparse mountain world. Mountain goats eat plants, grasses, mosses, and other alpine vegetation.

Odd-toed Ungulate

Odd-toed ungulate is the common name for any of the hoofed, herbivorous, terrestrial mammals comprising the order Perissodactyla, characterized by a pulley-like groove in the proximal surface of the astragalus (a bone in the ankle joint) and an odd number of functional toes (one or three), with the axis of each limb going through the enlarged middle (third) toe. Perissodactyls include such well-known members as horses, zebras, tapirs, and rhinoceroses. Perissodactyla is one of two living orders of ungulates (hoofed mammals), the other being Artiodactyla, the even-toed ungulates, which have a double-pulley astragalus and an even number (two or four) of functional toes.

The extant odd-toed ungulates are usually large, have relatively simple stomachs, and are exclusively herbivorous. In contrast to the ruminant Artiodactyl ungulates, perissodactyls are hindgut fermenters; that is, they digest plant cellulose in their intestines rather than in the first chamber (rumen) of the stomach.

Perissodactyls play key roles ecologically, commercially, and culturally. Ecologically, they provide a function in terrestrial food chains, although the large size and speed of the adults of many species, and the bristly manes of tapirs and horns of rhinoceroses, make them difficult prey as adults for all but the largest predators. Commercially and culturally, the domestication of some species, notably the horse and donkey, has been of great importance for human mobility, warfare, and transportation. Tapirs also are a popular food source and sought for sports hunting as well, and their skins provide leather. Rhinos are harvested, illegally, for their horns and other body parts for traditional Asian medicine. With the exception of the domesticated species, most perissodactyls are threatened with extinction.

The order Perissodactyla traditionally is divided into three extant families, six genera, and about 18 species. The Tapiridae family comprises one genera and four species of tapirs, the Rhinocerotidae includes four genera and five species of rhinoceroses, and the Equidae comprises one genera and nine species of horses, asses, and zebras.

Perissodactyls are characterized by their single functional toe, or three functional toes together, bearing the weight of the animals, with the axis of the limbs passing through the middle digit, which is enlarged. (This would be the third digit according to the five-toed ancestral plan.) Members of the family Equidae (horses, zebras, etc.) have one functional toe. Members of the Rhinocerotidae (rhinoceroses) have three digits on all four feet. Members of the Tapiridae (tapirs) have three digits on the hind feet, but four digits on the forefeet. All perissodactyls, extinct and extant, have this mesaxonic foot structure, meaning the symmetry of the foot passes through the third digit. Unlike the artiodactylas, the median metacarpals and metatarsal bones in the feet are not fused into a cannon bone.

All perissodactyls are hindgut fermenters. Hindgut fermenters, in contrast to the ruminants, store digested food that has left the stomach in a pouch-like extension of the large intestine called the caecum (literally "cave"), where the food is digested by bacteria. Food passes through the stomach twice as fast as the ruminant foregut fermenters and digestion is less efficient: the efficiency of digestion of the horse is only about seventy percent of that of cattle. Thus, perissodactyls consume more food per unit of weight than ruminants within the even-toed ungulates.

The living perissodactyls are a diverse group, with no generalized appearance. At one extreme are the lithe and graceful horses; on another, the huge, tank-like rhinoceroses; and in the middle, the vaguely pig-like tapirs. All extant perissodactyls are large, from the mountain tapir, which reaches 200 kilograms (485 pounds) to the white rhinoceros, which reaches to over 3500 kilograms (7,700 pounds).

Extinct perissodactyls possessed a far more diverse range of forms, including the tiny, vaguely tapir-like paleotheres, the monstrous brontotheres, the knuckle-walking chalicotheres, and the gigantic rhinoceros Indricotherium, which dwarfed even elephants.

Distribution and Habitat

Perissodactyls are limited to Africa, Asia, and the Americas. They were more widespread just a few hundred years ago, including in Europe, but the wild horses became extinct on that continent in the nineteenth century. They occupy a wide diversity of habitats, including deserts, rainforest, grasslands, and scrublnds.

Behavior, Diet and Reproduction

Today, the equines are the only social perissodactyls still extant. Horses organize themselves into small bands with a dominant mare at the top of the pecking order, as well as a resident stallion. Several bands will share a common territory, with some members of one band joining another band, every so often. These bands, in turn, form a "herd." Huge fossil beds made of the bones of hundreds or thousands of individuals suggest that many of the larger brontothere species, and some prehistoric rhinoceroses, such as Diceratherium were also social animals, which organized themselves into herds. Modern-day rhinoceroses, on the other hand, are solitary animals that maintain territories, often attacking members of their own species when their space has been invaded. Tapirs, too, are solitary animals, though they are shy, retiring creatures, which do not defend or maintain territories.

A tapir calf

All extant members of Perissodatyl are strictly herbivorous. Tapirs are browsers and frugivores, rhinos forage on leafy material, and woody and grassy vegetation and sometimes fruits, and equids forage on fibrous foods (largely grasses and sedges, but also leaves, buds, fruits, roots, etc.).

Perissodatyls are slow breeders, with a low reproductive rate. Perissodactyls tend to have one foal or calf at a time. Very rarely, the female may have twins. Gestation is very long, from about 11 months in horses to 16 months for rhinoceroses. The calf or foal is capable of standing within moments of birth, but is very dependent on its mother. The young stays with its mother even after weaned, usually until it is chased off by the mother upon the birth of a new foal or calf. At this time, in horses, the foal will enter into the herd proper, while young stallions are often chased off and join bachelor herds. With rhinos and tapirs, the newly weaned calf wanders away to search for new feeding grounds.

As with the males of many other animal groups, male perissodactyls often spar with each other for the privilege to mate with receptive females. A male which has found a female will attempt to taste her urine in order to see if she is in estrus. The female may also signal that she is in estrus, such as the whistling of cow Indian rhinoceroses and tapirs.

Humans and Conservation

Domestication

Humans have a historically long interaction with perissodactyls. The wild ass was the first equid to be domesticated, around 5000 B.C.E. in Egypt. Horses were domesticated 1000 years later, in the late neolithic. The original purpose of domesticating the horse may have been for food, but about 4,000 years ago they became a means of transport, a development that caused a revolution in human mobility and in warfare. During the sixteenth century, the Spaniards brought horses with them, and inadvertently reintroduced horses back into North America. While no rhinoceros has been domesticated, they have been captured for zoos and menageries since ancient times. The zebroid, that is, a zebra hybrid, began appearing in zoos and menageries during the nineteenth century.

In addition, wild species of perissodactyls have great importance as food and the skins of tapirs are used to provide good quality leather. Rhinoceroses are illegally hunted for horns and other body parts that are used in traditional Asian medicine.

Conservation

The Przewalski's horse, one of the most endangered equids

The odd-toed ungulates have been among the most important herbivorous mammals. At times, they have been the dominant herbivores in many ecosystems. However, over the course of millions of years, many species went extinct due to climatic change, newer, coarser-leaved plants, predators, disease, and competition from other herbivores, particularly the artiodactyls. The Chalicotheriidae was the most recent family of perissodactyl to become entirely extinct. The perissodactyls' decline continues even today. Most species are listed as threatened species, and although no species are confirmed to be extinct, some subspecies have gone extinct. The quagga was hunted for its meat, the tarpan were hunted for sport, and a subspecies of black rhinoceros was hunted for its horn (as with all other African rhinoceros species).

Perissodactyls tend to do well in captivity, and there are many breeding programs in place to help replenish wild populations. The Przewalski's horse has been recently released back to the wild. Some of the captive breeding programs for some equids are unusual, in that breeders have been carefully selecting specimens in order to recreate various recently extinct equids, such as the tarpan and quagga. Most wild rhinoceroses are monitored, and some have their horns trimmed off in order to discourage horn-poachers. Even so, if conservations do not improve, it may very well be that the only living perissodactyls left will be the domesticated horse and donkey.

Evolution

Although no certain records are known prior to the early Eocene, the odd-toed ungulates probably arose in what is now Asia during the late Paleocene, less than 10 million years after the Cretaceous-Tertiary extinction event, in which the dinosaurs (and all other large animals) died out (Hooker 2005). By the start of the Eocene (55 million years ago), they had diversified and spread out to occupy several continents. The horses and tapirs both arose in North America (Savage and Long 1986), and the rhinoceroses appear to have developed in Asia from tapir-like animals and then recolonized the Americas during the middle Eocene (about 45 million years ago). About 15 families have been recognized, of which only three survive (McKenna and Bell 1997; Hooker 2005). These 15 families were very diverse in form and size; they included the enormous brontotheres and the bizarre chalicotheres. The largest perissodactyl, an Asian rhinoceros called Paraceratherium, reached 11 metric tons (12 short tons), more than twice the weight of an elephant.

Perissodactyls were the dominant group of large terrestrial browsers right through the Oligocene. However, the rise of grasses in the Miocene (about 20 million years ago) saw a major change: the even-toed ungulates soon rose to prominence, which is probably a function of their more complex stomachs, which were better able to adapt to a coarse, low-nutrition diet. Nevertheless, many odd-toed species survived and prospered until the late Pleistocene (about 10,000 years ago) when they faced the pressure of human hunting and habitat change.

Classification

The members of the order fall into two suborders:

- Hippomorpha are odd-toed ungulates that are, today, fast runners with long legs and have only one toe. The only extant family of this suborder is Equidae (whose sole surviving genus is Equus), comprising the horse, zebra, donkey, onager, and allied species. The extinct, rhinoceros-like brontotheres are also included in this suborder. Both families probably descended from palaeotheres.
- Ceratomorpha have several functional toes; they are heavier than and move slower than the Hippomorpha. This suborder has two extant families: Tapiridae (tapirs) and Rhinocerotidae (rhinoceroses). The extinct chalicotheres may belong to this suborder as well.

The three surviving families of odd-toed ungulate are classified as follows.

- ORDER PERISSODACTYLA
  - Suborder Hippomorpha
    - Family Equidae: horses and allies, nine species in one genus
      - Przewalski's Horse, Equus przewalskii
      - Domestic Horse, Equus caballus
      - Wild horse, Equus ferus
      - Donkey or African Ass, Equus asinus
      - Onager or Asiatic Ass, Equus hemionus
      - Kiang or Tibetan Wild Ass, Equus kiang
      - Plains Zebra, Equus quagga
      - Cape Mountain Zebra, Equus zebra

        - Hartmann's Mountain Zebra, Equus hartmannae
        - Grevy's Zebra, Equus grevyi
  - Suborder Ceratomorpha
    - Family Tapiridae: tapirs, four species in one genus
        - Brazilian Tapir, Tapirus terrestris
        - Mountain Tapir, Tapirus pinchaque
        - Baird's Tapir, Tapirus bairdii
        - Malayan Tapir, Tapirus indicus
    - Family Rhinocerotidae: rhinoceroses, five species in four genera
        - Black Rhinoceros, Diceros bicornis
        - White Rhinoceros, Ceratotherium simum
        - Indian Rhinoceros, Rhinoceros unicornis
        - Javan Rhinoceros, Rhinoceros sondaicus
        - Sumatran Rhinoceros, Dicerorhinus sumatrensis

It was thought that odd-toed ungulates are closely related to even-toed ungulates. But recent molecular comparisons show that even-toed and odd-toed ungulates may not form a clade. Instead, perissodactyls may be more closely related to carnivorans, bats, and pangolins (and, by default, the Creodonts and Cimolestids). So, some biologists group the orders Perissodactyla, Carnivora, Chiroptera, Pholidota, Creodonta, and Cimolesta as the clade Pegasoferae.

Two Recently Extinct Equids

The quagga became extinct

The tarpan became extinct

Rhinoceros

The rhinoceros is a large mammal native to Africa and Asia. There are five species of rhino found in the world with 3 out of the 5 species of rhino now considered to be critically endangered. The rhino is thought to be the second biggest land mammal in the world behind the African elephant.

The five species of rhinoceros are the white rhino (which is the largest species of rhino) and the black rhino which are both native to Africa and are only really distinguished in size as they look fairly similar. The Indian rhino, the Sumatran rhino and the Javan rhino are all native to Asia and are much smaller in size than the white rhino and the black rhino of Africa.

Rhinoceroses are characterized by the possession of one or two horns on the upper surface of the snout; these horns are not true horns but are composed of keratin, a fibrous protein found in hair. Modern rhinoceroses are large animals, ranging from 2.5 metres (8 feet) long and 1.5 metres (5 feet) high at the shoulder in the Sumatran rhinoceros to about 4 metres (13 feet) long and nearly 2 metres (7 feet) high in the white rhinoceros. Adults of larger species weigh 3–5 tons. Rhinoceroses are noted for their thick skin, which forms platelike folds, especially at the shoulders and thighs. All rhinos are gray or brown in colour, including the white rhinoceros, which tends to be paler than the others. Aside from the Sumatran rhinoceros, they are nearly or completely hairless, except for the tail tip and ear fringes, but some fossil species were covered with dense fur. The feet of the modern species have three short toes, tipped with broad, blunt nails.

Most rhinoceroses are solitary. Individuals usually avoid each other, but the white rhinoceros lives in groups of up to 10 animals. In solitary species the home territory is crisscrossed with well-worn trails and often marked at the borders with urine and piles of dung.

Northen White rhinoceros

Rhinoceroses have poor eyesight but acute senses of hearing and smell. Most prefer to avoid humans, but males, and females with calves, may charge with little provocation. The black rhinoceros (Diceros bicornis) is normally ill-tempered and unpredictable and may charge any unfamiliar sound or smell. Despite their bulk, rhinoceroses are remarkably agile; the black rhinoceros can attain a speed of about 45 km (30 miles) per hour, even in thick brush, and can turn around rapidly after missing a charge. Like elephants, rhinoceroses communicate using infrasonic frequencies that are below the threshold of human hearing. The use of infrasonic frequencies is likely an adaptation for rhinoceroses to keep in touch with each other where they inhabit dense vegetation and probably for females to advertise to males when females are receptive to breeding.

Black rhinoceroses(Diceros bicornis) in a Zoo

Rhinoceroses are by far the largest of the perissodactyls, an order of hoofed mammals that also includes the horses and zebras. One of the features of very large body size in mammals is a low reproductive rate. In rhinoceroses, females do not conceive until about six years of age; gestation is long (16 months in most species), and they give birth to only one calf at a time. The period of birth between calves can range from 2 to 4.5 years. Thus, the loss of a number of breeding-age females to poachers can greatly slow the recovery of rhinoceros populations. However, an Indian rhinoceros female will conceive again quickly if she loses her calf. In this species tigers kill about 10–20 percent of calves. Tigers rarely kill calves older than 1 year, so those Indian rhinoceroses that survive past that point are invulnerable to nonhuman predators.

White rhinoceroses

The three Asian species fight with their razor-sharp lower outer incisor teeth, not with their horns. In Indian rhinoceroses such teeth, or tusks, can reach 13 cm (5 inches) in length among dominant males and inflict lethal wounds on other males competing for access to breeding females. The African species, in contrast, lack these long tusklike incisors and instead fight with their horns.

The rhinoceros's horn is also the cause of its demise. Powdered rhinoceros horn has been a highly sought commodity in traditional Chinese medicine—not as an aphrodisiac, as is often widely reported, but as an antifever agent. Substitute agents have been found, particularly pig bone and water buffalo horn, but rhinoceros horn commands tens of thousands of dollars per kilogram in Asian markets. Today poaching remains a serious problem throughout the range of all species of rhinoceros.

The term rhinoceros is sometimes also applied to other, extinct members of the family Rhinocerotidae, a diverse group that includes several dozen fossil genera, among them the woolly rhinoceros (Coelodonta antiquitatis). Early rhinoceroses resembled small horses and lacked horns. (Horns are a relatively recent development in the lineage.) The largest land mammal ever to have lived was not an elephant but Indricotherium, a perissodactyl that was 6 metres (20 feet) long and could browse treetops like a giraffe.

White Rhinoceros

White rhinos are the second largest odd-toed ungulate land mammal and their name comes from the Afrikaan's, a West Germanic language, word "weit" which means wide and refers to the animal's mouth. Also known as the square-lipped rhinoceros, white rhinos have a square upper lip with almost no hair. Two genetically different subspecies exist, the nothnern and southern white rhino and are found in two different regions in Africa. As of March 2018, there are only two rhinos of the northern white rhino left, both of which are female. They live in the Ol Pejeta Conservancy in Kenya and are protected round-the-clock by armed guards. Their near extinction is due to decades of rampant poaching for rhino horn.

White rhinos have complex social structures. Groups of sometimes 14 rhinos may form, notably females with calves. Adult males defend territories of roughly one square mile, which they mark with vigorously scraped dung piles. The home range for adult females can be more than seven times larger, depending on habitat quality and population density. Breeding females are prevented from leaving a dominant male's territory, which is marked and patrolled by its owner on a regular basis. Males competing for a female may engage in serious conflict, using their horns and massive size to inflict wounds.

Description

Comparative illustration of black (top) and white rhinos (bottom)

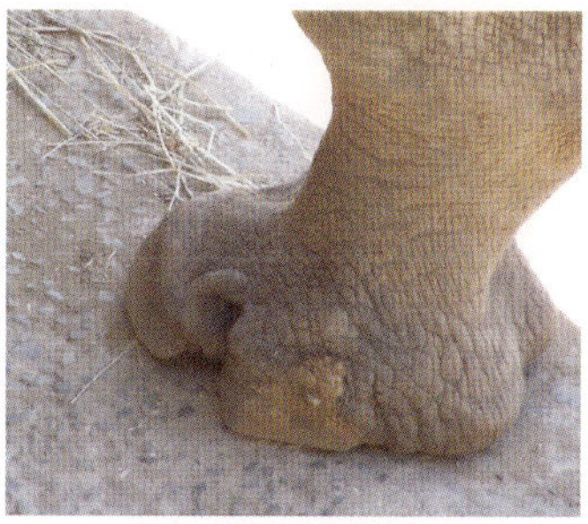

White rhinos have three distinct toes

The white rhinoceros is the largest of the five living species of rhinoceros. By mean body mass, the white rhinoceros falls behind only the three extant species of elephant as the largest land animal and terrestrial mammal alive today. It weighs slightly more on average than a hippopotamus despite a considerable mass overlap between these two species. It has a massive body and large head, a short neck and broad chest. The head and body length is 3.7 to 4.5 m (12.1 to 14.8 ft) in males and 3.4 to 3.65 m (11.2 to 12.0 ft) in females, with the tail adding another 70 cm (28 in) and the shoulder height is 170 to 186 cm (5.58 to 6.10 ft) in the male and 160 to 177 cm (5.25 to 5.81 ft) in the female. The male, averaging about 2,300 kg (5,070 lb) is heavier than the female, at an average of about 1,700 kg (3,750 lb). The largest size the species can attain is not definitively known; specimens of up to 3,600 kg (7,940 lb) are considered reliable, while larger sizes up to 4,500 kg (9,920 lb) have been claimed but are not verified. On its snout it has two horn-like growths, one behind the other. These are made of solid keratin, in which they differ from the horns of bovids (cattle and their relatives), which are keratin with a bony core, and deer antlers, which are solid bone.

The skull

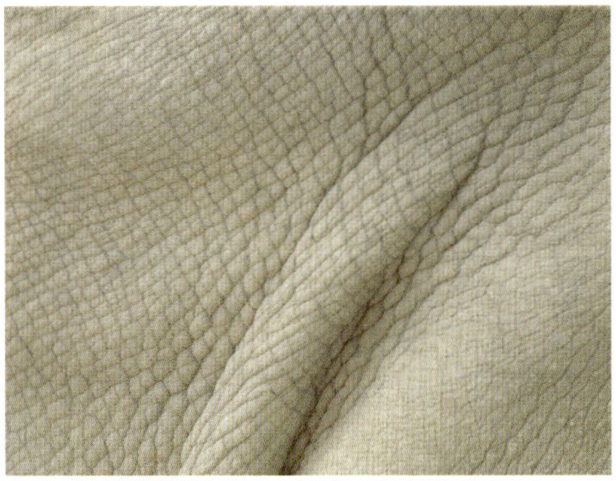

Closeup of skin at Lisbon Zoo

The front horn is larger and averages 60 cm (24 in) in length, reaching as much as 150 cm (59 in) but only in females. The white rhinoceros also has a noticeable hump on the back of its neck. Each of the four stumpy feet has three toes. The color of the body ranges from yellowish brown to slate grey. Its only hair is the ear fringes and tail bristles. White rhinos have a distinctive broad, straight mouth which is used for grazing. Its ears can move independently to pick up sounds, but it depends most of all on its sense of smell. The olfactory passages that are responsible for smell are larger than their entire brain. The white rhinoceros has the widest set of nostrils of any land-based animal.

Behavior and Ecology

White rhinoceroses are found in grassland and savannah habitat. Herbivore grazers that eat grass, preferring the shortest grains, the white rhinoceros is one of the largest pure grazers. It drinks twice a day if water is available, but if conditions are dry it can live four or five days without water. It spends about half of the day eating, one third resting, and the rest of the day doing various other things. White rhinoceroses, like all species of rhinoceros, love wallowing in mud holes to cool down. The white rhinoceros is thought to have changed the structure and ecology of the savanna's grasslands. Comparatively based on studies of the African elephant, scientists believe the white rhino is a driving factor in its ecosystem. The destruction of the mega herbivore could have serious cascading effects on the ecosystem and harm other animals.

White rhinoceros wallowing in the mud

White rhinoceroses produce sounds which include a panting contact call, grunts and snorts during courtship, squeals of distress, and deep bellows or growls when threatened. Threat displays (in males mostly) include wiping its horn on the ground and a head-low posture with ears back, combined with snarl threats and shrieking if attacked. The vocalisations of the two species are slightly different. The white rhinoceros is quick and agile and can run 50 km/h (31 mph).

White rhinoceroses live in crashes or herds of up to 14 animals (usually mostly female). Sub-adult males will congregate, often in association with an adult female. Most adult bulls are solitary. Dominant bulls mark their territory with excrement and urine. The dung is laid in well-defined piles. It may have 20 to 30 of these piles to alert passing rhinoceroses that it is his territory. Another way of marking their territory is wiping their horns on bushes or the ground and scrapes with its feet before urine spraying. They do this around ten times an hour while patrolling territory. The same ritual as urine marking except without spraying is also commonly used. The territorial male will scrape-mark every 30 m (98 ft) or so around its territory boundary. Subordinate males do not mark territory. The most serious fights break out over mating rights with a female. Female territory overlaps extensively, and they do not defend it.

Reproduction

A calf of white rhinoceros

Young rhino with mother at Rhino and Lion Nature Reserve, Johannesburg

Females reach sexual maturity at 6–7 years of age while males reach sexual maturity between 10–12 years of age. Courtship is often a difficult affair. The male stays beyond the point where the female acts aggressively and will give out a call when approaching her. The male chases and or blocks the way of the female while squealing or wailing loudly if the female tries to leave his territory. When ready to mate the female curls her tail and gets into a stiff stance during the half-hour copulation. Breeding pairs stay together between 5–20 days before they part their separate ways. The gestation period of a white rhino is 16 months. A single calf is born and usually weighs between 40 and 65 kg (88 and 143 lb). Calves are unsteady for their first two to three days of life. When threatened, the baby will run in front of the mother, which is very protective of her calf and will fight for it vigorously. Weaning starts at two months, but the calf may continue suckling for over 12 months. The birth interval for the white rhino is between two and three years. Before giving birth, the mother will chase off her current calf. White rhinos can live to be up to 40–50 years old.

Adult white rhinos have no natural predators (other than humans) due to their size, and even young rhinos are rarely attacked or preyed on due to the mother's presence and their tough skin. One exceptional successful attack was perpetrated by a lion pride on a roughly half-grown white rhinoceros, which weighed 1,055 kg (2,326 lb), and occurred in Mala Mala Game Reserve, South Africa.

Distribution

The southern white rhino lives in Southern Africa. About 98.5% of white rhinos live in just five countries (South Africa, Namibia, Zimbabwe, Kenya and Uganda). Almost at the edge of extinction in the early 20th century, the southern subspecies has made a tremendous comeback. In 2001 it was estimated that there were 11,670 white rhinos in the wild with a further 777 in captivity worldwide, making it the most common rhino in the world. By the end of 2007 wild-living southern white rhino had increased to an estimated 17,480 animals.

The northern white rhino (*Ceratotherium simum cottoni*) formerly ranged over parts of north-western Uganda, southern Chad, south-western Sudan, the eastern part of Central African Republic, and north-eastern Democratic Republic of the Congo (DRC). The last surviving population of wild northern white rhinos are or were in Garamba National Park, Democratic Republic of the Congo (DRC) but in August 2005, ground and aerial surveys conducted under the direction of African

Parks Foundation and the African Rhino Specialist Group (ARSG) have only found four animals: a solitary adult male and a group of one adult male and two adult females. In June 2008 it was reported that the species may have gone extinct in the wild.

Like the black rhino, the white rhino is under threat from habitat loss and poaching, most recently by Janjaweed. Although there are no measurable health benefits, the horn is sought after for traditional medicine and jewelry.

Poaching

A female southern white rhino shot

Historically the major factor in the decline of white rhinos was uncontrolled hunting in the colonial era, but now poaching for their horn is the primary threat. The white rhino is particularly vulnerable to hunting, because it is a large and relatively unaggressive animal with very poor eyesight and generally occurs in herds.

Despite the lack of scientific evidence, the rhino horn is highly prized in traditional Asian medicine, where it is ground into a fine powder or manufactured into tablets to be used as a treatment for a variety of illnesses such as nosebleeds, strokes, convulsions, and fevers. Due to this demand, several highly organized and very profitable international poaching syndicates came into being and would carry out their poaching missions with advanced technologies ranging from night vision scopes, silenced weapons, darting equipment and even helicopters. The ongoing civil war in the Democratic Republic of Congo and incursions by poachers primarily coming from Sudan have further disrupted efforts to protect the few remaining northern rhinos.

In 2013, poaching rates for white rhinos nearly doubled from the previous year. As a result, the white rhino has now received Near Threatened status as its total population tops out at 20,000 members. Poaching of the animal has gone virtually unchecked in most of Africa, and the non-violent nature of the rhinoceros makes it susceptible to poaching. Mozambique, one of the four main countries the white rhino occupies, is used by poachers as a passageway to South Africa, which holds a fairly large number of white rhinos. Here, rhinos are regularly killed and their horns are smuggled out of the country. As of 2014, Mozambique labels white rhino poaching as a misdemeanor.

In March, 2017 the Thoiry Zoo, located in France, was broken into by poachers. A Southern white rhinoceros named Vince was found shot dead in his enclosure; the poachers had removed one of his horns and had attempted to remove his second horn. This is believed to be the first time a rhinoceros had been killed in a European zoo.

Even with increased anti-poaching efforts in many African countries, many poachers are still willing to risk death or prison time because of the tremendous amount of money that they stand to make. Rhino horn can fetch tens of thousands of dollars per kilogram on the black market in Asia and, depending on the exact price, can be worth more than its weight in gold. Poachers are also starting to use social media sites for obtaining information on the location of rhino in popular tourist attractions (such as Kruger National Park) by searching for geotagged photographs posted online by unsuspecting tourists. By using GPS coordinates of rhino in recent photographs, poachers are able to more easily find and kill their targets.

Shire Horse

Shire horse breed native to the middle section of England. Shire horses are known as the tallest breed of draft horse. The stallions average a colossal 17 hands, or 68 inches at the shoulder, with mares a bit smaller. Shire horses are frequent record holders for height. In comparison, riding horses like Thoroughbreds and quarter horses average between 15 and 16 hands tall, or 60 to 64 inches at the shoulder, and weigh around 1,100 pounds. Shire horses often weigh around 2,000 pounds.

Shires have the strengths and weaknesses of any draft breed, with some issues particularly affecting them.

Work Ethic

If you're looking for a working horse, it's hard to beat the shire for strength and work ethic. He learns easily and wants to please. This draft horse, primarily serving inr driving or plowing, is suited to training as a riding horse. Properly conditioned, he can work long hours in the field, on the road or under saddle.

Temperament

Shires generally like humans and possess good temperaments. Such large horses must be docile, as they would otherwise be quite dangerous to handle. Shires possess patience and don't spook easily. This is the breed that pulled omnibuses in major cities before the advent of the automobile. Not much surprises them, and they can deal with the kind of stimuli that would cause a meltdown in less calm equines.

Costs

Draft horses such as the shire cost more to feed and shoe than "light" horses -- often a lot more. At maturity, a shire weighs 2,000 pounds or more. A light horse may weigh half of that. In fact, the shire is the largest of the draft breeds, according to the Rare Breeds Trust of Australia. The breed standard requires a minimum height of 16.2 hands full-grown; many reach 19 hands or more. Because of his size, you might need to purchase custom-made tack for your shire. Their relative rarity also makes a shire more expensive than more common horses.

Rarity

Although the number of registered shires is increasing, the breed almost became extinct in the

United States. According to the American Livestock Breeds Conservancy, the U.S. shire registry ceased operations from 1955 to 1961, but after that a slow but steady revival took place. If you want to purchase a shire, it could take more time and travel than with more common draft breeds to find the right animal.

Health Issues

Like other draft breeds, shires are prone to polysaccharide storage myopathy, a neuromuscular disease that causes hind leg spasms. Your vet might recommend a high-fat diet to avoid this problem. The shire's heavily feathered legs require a great deal of upkeep to keep clean. Because of the feathering, shires often suffer from the bacterial infection colloquially known as "mud fever," "scratches" and "dew poisoning." While topical treatment or antibiotics can help, keeping your shire out of wet areas reduces the odds of his developing this dermatological disease.

The Shire horse was used extensively to pull carts of ale from breweries to public houses. Before World War I, these gentle giants commonly served as farm horses, pulling wagons and working agricultural fields. When coal was a major source of heat and light, Shires were employed to pull enormous coal wagons over the rough roads.

Today, the Shire is still popular for pulling vehicles such as sightseeing wagons, and many equestrians enjoy riding the gentle breed for pleasure. They are an environmentally friendly alternative to tractors on small farms and are still used in logging operations. Lastly, but perhaps most visibly, they are a popular promotional tool for modern-day beer brewers, some of whom are again making deliveries by horse and wagon.

Champion and Celebrity Shire Horses

Great Britain's tallest horse is Duke, a shire who, as of 2008, stood at a regal 19.3 hands high (that's about 6 feet 5 inches), weighed about the same as a car, and was still growing. He eats about four times the amount average-sized horses consume and lives in a stable that was built especially for him (he can't fit into standard-sized stables). His blankets and other accessories are, of necessity, custom-made. Despite his size, though, his handlers say he is "as gentle as a lamb" and is very attached to a Shetland pony that lives on the same farm.

One of the most famous Shires was the majestic Goliath, who died in 2001. Measured at more than 19.5 hands high, he was the tallest horse in Great Britain at the time and, for eight years, was listed in "The Guinness Book of World Records." In fact, the publication recognized him in 2000 as the tallest living horse in the world. A beautiful black horse renowned for his character, he was a member of a show horse team owned by the Young and Co. brewery.

References

- Estes, R. (1992). The Behavior Guide to African Mammals: including hoofed mammals, carnivores, primates. University of California Press. pp. 222–26. ISBN 0-520-08085-8
- Horwitz, Liora Kolska; Eitan Tchernov (1990). "Cultural and Environmental Implications of Hippopotamus Bone Remains in Archaeological Contexts in the Levant". Bulletin of the American Schools of Oriental Research. 280: 67–76. doi:10.2307/1357310
- Strengths-weaknesses-shire-horse-3951: animals.mom.me, Retrieved 18 May 2018

- "Hippopotamus amphibius". National Geographic. Archived from the original on 25 November 2014. Retrieved 10 July 2013
- William E. Barklow (2004). «Low-frequency sounds and amphibious communication in Hippopotamus amphibious». The Journal of the Acoustical Society of America. 115 (5): 2555. doi:10.1121/1.4783854
- Learn-about-the-shire-horse-breed-1886127: thesprucepets.com, Retrieved 28 March 2018
- Kingdon, J. (1988). East African Mammals: An Atlas of Evolution in Africa, Volume 3, Part B: Large Mammals. University Of Chicago Press. pp. 256–77. ISBN 0-226-43722-1
- Haas, Georg (1953). "On the Occurrence of Hippopotamus in the Iron Age of the Coastal Area of Israel". Bulletin of the American Schools of Oriental Research. 132: 30–34. doi:10.2307/1355798
- Even-toed-ungulate: newworldencyclopedia.org, Retrieved 30 June 2018
- Owen, James (24 October 2006). "Hippos Butchered by the Hundreds in Congo Wildlife Park". National Geographic News. Retrieved 11 September 2013
- McCarthy, T.S.; Ellery, W. N.; Bloem, A (1998). "Some observations on the geomorphological impact of hippopotamus (Hippopotamus amphibius L.) in the Okavango Delta, Botswana". African Journal of Ecology. 36 (1): 44–56. doi:10.1046/j.1365-2028.1998.89-89089.x

Chapter 4

Aquatic Mammals

Aquatic mammals are the mammals that dwell in water bodies. Examples of such mammals are the blue whale, Amazon river dolphin, elephant seal, walrus, etc. All the significant kinds of aquatic mammals inhabiting the seas, oceans and freshwater bodies of the world have been carefully analyzed in this chapter.

Aquatic and semiaquatic mammals are well adapted to life in the water. The aquatic mammals have evolved from terrestrial mammals. The fact that all of them are not gill-breathers but breathe air through lungs, indicate their original terrestrial mode of life.

All the aquatic mammals are really terrestrial lung-breathing forms which have reverted to an aquatic life, and they have done so with remarkable success, the whales being the most successful. They have reverted to water probably because of extreme competition on land for food and shelter.

There are several aquatic mammals. Aquatic mammals belong to several orders of Mammalia.

Depending on the degree for aquatic adaptation the aquatic mammals have been divided into the following categories:

1. Amphibious mammals

These mammals do not live permanently in water. They live on land but go into water for food and shelter.

They show only partial aquatic adaptations such as:

1. Small external ears,
2. Webbed feet,
3. Flattened nails,
4. Subcutaneous fat.

The mammals of this category include the beaver (Castor), musk rat (Ondatra), nutria (Myocaster), otter (Lutra), mink (Mustela) and many others. The amphibious mammals belong to several orders of mammalia such as Carnivora, Rodentia, Artiodactyla, Marsupialia, Monotremata, etc.

2. Aquatic mammals

The mammals under this category spend most of the time in water and usually come to land for reproduction. The typical examples are seals and hippopotamus.

3. Marine mammals

These mammals never come to land and are perfectly at home in water. The typical examples are whales.

Aquatic Adaptations

The adaptations or specialisations of truly aquatic mammals (Cetacea and Sirenia) are divided into 3 major categories:

1. Modifications of original structures,
2. Loss of structures, and
3. Development of new structures.

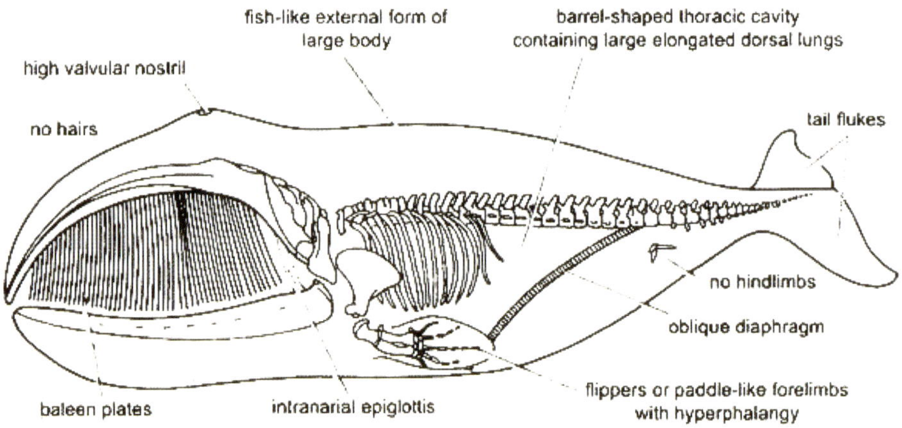

Figure: Body outline and skeleton of Greenland right whale *Balaena mysticetus* showing aquatic adaptaions.

A. Modifications of Original Structures

1. Body shape

In aquatic mammals, body shape is of prime importance. The external fishlike form, elongated head, indistinct neck and tapering streamlined body offers little resistance and swims rapidly through water.

2. Large size and weight

In aquatic mammals, the large size and body weight help the aquatic mammals. Whalebone whale may grow up to 35 metres in length and weigh about 150 tons. Large size reduces skin friction and loss of heat, but creates no problem for support in water due to buoyancy.

3. Flippers

In aquatic mammals, the forelimbs are transformed into skin-covered, un-jointed paddles or flippers, having no separate indication of fingers. These paddles or flippers can move as a whole only at the shoulder joint. The broad and flattened paddles or flippers serve as balancers and provide stability during swimming.

4. Hyperdactyly and hyperphalangy

In aquatic mammals, extra digits (hyperdactyly) and extra phalanges (hyperphalangy) up to 14 or more in some forms, serve to increase the surface area of flippers for greater utility for swimming in water.

5. High and valvular nostrils

In aquatic mammals, the nostrils are placed far back on the top of head so that animal can breathe air without raising head much out of water. The nostrils can also be closed by valves during diving under water.

6. Mammary ducts

In aquatic mammals, during lactation, ducts of mammary glands dilate to form large reservoirs of milk, which is pumped directly into mouth of young by the action of special compressor muscle. This arrangement facilitates suckling of young under water.

7. Oblique diaphragm

In aquatic mammals, oblique diaphragm makes the thoracic cavity larger dorsal and barrel-shaped for providing more space to lungs for expansion.

8. Large lungs

In aquatic mammals, large unlobulated and highly elastic lungs ensure taking down maximum air Lore submergence. Like swim bladders of fishes, the dorsal lungs also serve as hydrostatic organs in maintaining a horizontal posture during swimming.

9. Intra-narial epiglottis

In aquatic mammals, elongated, tubular and intra-narial epiglottis, when embraced by the soft palate, provides a continuous and separate air-passage, thus, allowing breathing and feeding simultaneously.

10. Endoskeleton

In aquatic mammals, the cranium becomes small but wider to accommodate the short and wide brain. The facial part of skull projects forming elongated snout or rostrum

The zygomatic arches are reduced. Due to reduced neck, the cervical vertebrae are fused into a solid bony mass. Zygapophyses are reduced. Sacrum is also reduced. Ribs become arched dorsally to increase thoracic cavity. Bones are light and spongy. In Cetacea, bones are filled with oil.

11. Teeth

In toothed whales, teeth are monophyodont, homodont and numerous, as many as 250. This helps in capturing or seizing prey, prevent its escape and swallowing it without mastication. Usually, the mobility of jaws is reduced as they have no function in mastication.

B. Loss of Structures

In aquatic mammals, there is a loss of a few structures which are usually present in other mammals.

These are as follows:

1. There is a loss of hairs. Skin surface usually remains smooth and glistening due to loss of hairs except for a few sensory bristles on snout or lips in some cases.
2. Pinnae are also absent. Presence of hairs and pinnae may obstruct or impede the ever flow of water over body surface and interfere with the speed and elegance of movement through water.
3. Nictitating membranes, eye cleansing glands, lacrimal glands and all kinds of skin glands (sweat and sebaceous) are also absent because they would have been useless under water.
4. Skin losses its muscles and nerves due to thickening and immobility.
5. Hind limbs are represented only by button-like knobs in the foetus but disappear in the adult.
6. Pelvis is also rudimentary.
7. Fingernails are absent except for traces in foetus.
8. Scrotal sacs are also absent and testes remain inside abdomen.

C. Development of New Structures

1. Tail flukes

In aquatic mammals, some large, lateral or horizontal expansions of the skin develop on tail. These expansions are called tail flukes. These are not supported by fin-ray. Their up and down strokes not only propel the body through water but enable rapid return to the surface for breathing after prolonged submersion.

2. Dorsal fin

In most Cetacea develop an unpaired adipose dorsal fin without internal skeletal support. It serves as a rudder or keel during swimming.

3. Blubber

In aquatic mammals, the blubber is the thick subcutaneous layer of fat, which compensates for the lack of hairy covering. Blubber acts as a heat insulator. It not only retains the warmth of the body but also provides a ready reservoir of food and water during emergency.

The fat also reduces the specific gravity of the animal, thus, imparting buoyancy. Blubber also provides an elastic covering to allow changes in body volume during deep diving and also counteracts the hydrostatic pressure.

4. Baleen

In whalebone whales, teeth are absent. Instead, the upper jaw carries two transverse rows of numerous triangular fringed horny plates of baleen or whalebone. These serve as an effective sieve for straining plankton (mostly kril) which forms their chief food.

5. Foam

Foam is a fine emulsion of fat, mucus and gas. Each middle ear cavity sends an inner pneumatic prolongation, which meets with the fellow on the other side below the skull. These extensions contain foam. It probably serves to insulate sound and improves audition or hearing under water.

6. Melon

In some aquatic mammals, the melon is a receptor present in front of nostrils It consists of a fatty mass traversed by muscle fibres. It possibly serves to detect pressure changes in water.

7. Harderian glands

In aquatic mammals, eyes under water remain protected by a special fatty secretion of Harderian glands.

Ecology

Keystone Species

Beaver dams restrict water-flow, creating a pond

Beaver ponds have a profound effect on the surrounding ecosystem. Their first and foremost ecological function is as a reservoir for times of drought, and prevent drying of riverbeds. In the event of a flood, beaver ponds slow down water-flow which reduces erosion on the surrounding soil. Beaver dams hold sediment, which reduces turbidity and thereby improving overall water quality downstream. This supplies other animals with cleaner drinking water and prevents degradation of spawning grounds for fish. However, the slower water speed and lack of shade from trees (that have since been cut down to construct the dam), results in the overall temperature increasing. They also house predatory zooplankton which help break down detritus and control algae populations.

Diet

Beavers are herbivores, and prefer the wood of quaking aspen, cottonwood, willow, alder, birch, maple and cherry trees. They also eat sedges, pondweed, and water lilies. Beavers do not hibernate, but rather they store sticks and logs in a pile in their ponds, eating the under bark. The dams they build flood areas of surrounding forest, giving the beaver safe access to an important food supply, which is the leaves, buds, and inner bark of growing trees. They prefer aspen and poplar, but will also take birch, maple, willow, alder, black cherry, red oak, beech, ash, hornbeam and occasionally pine and spruce. They will also eat cattails, water lilies and other aquatic vegetation, especially in the early spring.

Indian rhinoceros are grazers. Their diets consist almost entirely of grasses, but they also eat leaves, branches of shrubs and trees, fruits, and submerged and floating aquatic plants. They feed in the mornings and evenings. They use their prehensile lips to grasp grass stems, bend the stem down, bite off the top, and then eat the grass. They tackle very tall grasses or saplings by walking over the plant, with legs on both sides and using the weight of their bodies to push the end of the plant down to the level of the mouth.

Manatees make seasonal movements synchronized with the flood regime of the Amazon Basin. They are found in flooded forests and meadows during the flood season when food is abundant, and move to deep lakes during the dry season. The Amazonian manatee has the smallest degree of rostral deflection (25° to 41°) among sirenians, an adaptation to feed closer to the water surface.

A moose in Siberia feeding on aquatic plants

A moose's diet often depends on its location, but they seem to prefer the new growths from deciduous trees with a high sugar content, such as white birch, trembling aspen and striped maple, among many others. They also eat many aquatic plants such as lilies and water milfoil. To reach high branches, a moose may bend small saplings down, using its prehensile lip, mouth or body. For larger trees a moose may stand erect and walk upright on its hind legs, allowing it to reach plants 14.0 feet (4.26 m) off the ground. Moose are excellent swimmers and are known to wade into water to eat aquatic plants. Moose are thus attracted to marshes and river banks during warmer months as both provide suitable vegetation to eat and water to bathe in. Moose have been known to dive underwater to reach plants on lake bottoms, and the complex snout may assist the moose in this type of feeding. Moose are the only deer that are capable of feeding underwater.

Hippopotamuses leave the water at dusk and travel inland, sometimes up to 10 km (6 mi), to graze on short grasses, their main source of food. They spend four to five hours grazing and can consume 68 kg (150 lb) of grass each night. Like almost any herbivore, they consume other plants if presented with them, but their diet consists almost entirely of grass, with only minimal consumption of aquatic plants. The pygmy hippopotamus emerges from the water at dusk to feed. It relies on game trails to travel through dense forest vegetation. It marks trails by vigorously waving its tail while defecating to further spread its feces. The pygmy hippo spends about six hours a day foraging for food, and they do not eat aquatic vegetation to a significant extent and rarely eat grass because it is uncommon in the thick forests they inhabit. The bulk of a pygmy hippo's diet consists of ferns, broad-leaved plants and fruits that have fallen to the forest floor. The wide variety of plants pygmy hippos have been observed eating suggests that they will eat any plants available. This diet is of higher quality than that of the common hippopotamus.

The Amazon river dolphin has the most diverse diet among cetaceans, consisting of at least 53 species of fish. They mainly feed on croakers, cichlids, tetras, and piranhas, but they may also target freshwater crabs and river turtles. South Asian river dolphins mainly eat fish (such as carp, catfish, and freshwater sharks) and invertebrates, mainly prawns.

Generally, all aquatic desmans, shrews, and voles make quick dives and catch small fish and invertebrates. The giant otter shrew, for example, makes quick dives that last for seconds and grabs small crabs (usually no bigger than 2.8 inches (7 cm) across). The Lutrine opossum is the most carnivorous opossum, usually consuming small birds, rodents, and invertebrates. Water voles mainly eat grass and plants near the water and at times, they will also consume fruits, bulbs, twigs, buds, and roots. However, a population of water voles living in Wiltshire and Lincolnshire, England have started eating frogs' legs and discarding the bodies.

Interactions with Humans

Exploitation

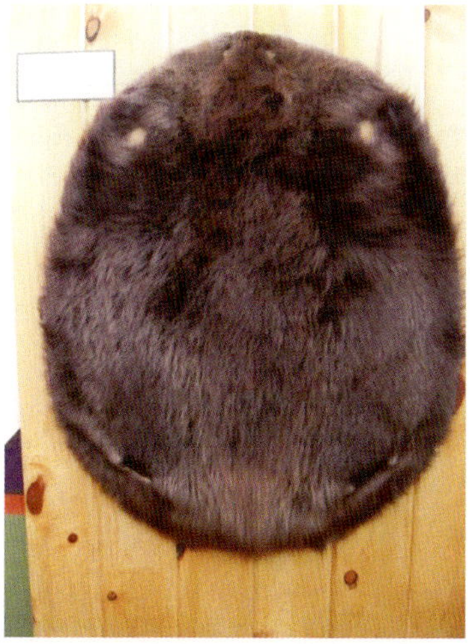

A beaver pelt in the Fur Trade Museum

Fur robes were blankets of sewn-together, native-tanned, beaver pelts. The pelts were called *castor gras* in French and "coat beaver" in English, and were soon recognized by the newly developed felt-hat making industry as particularly useful for felting. Some historians, seeking to explain the term *castor gras*, have assumed that coat beaver was rich in human oils from having been worn so long (much of the top-hair was worn away through usage, exposing the valuable under-wool), and that this is what made it attractive to the hatters. This seems unlikely, since grease interferes with the felting of wool, rather than enhancing it. By the 1580s, beaver "wool" was the major starting material of the French felt-hatters. Hat makers began to use it in England soon after, particularly after Huguenot refugees brought their skills and tastes with them from France.

Sport hunting of the Indian rhinoceros became common in the late 1800s and early 1900s. Indian rhinos were hunted relentlessly and persistently. Reports from the middle of the 19th century claim that some British military officers in Assam individually shot more than 200 rhinos. By 1908, the population in Kaziranga had decreased to around 12 individuals. In the early 1900s, the species had declined to near extinction. Poaching for rhinoceros horn became the single most important reason for the decline of the Indian rhino after conservation measures were put in place from the beginning of the 20th century, when legal hunting ended. From 1980 to 1993, 692 rhinos were poached in India. In India's Laokhowa Wildlife Sanctuary, 41 rhinos were killed in 1983, virtually the entire population of the sanctuary. By the mid-1990s, poaching had rendered the species extinct there. In 1950, Chitwan's forest and grasslands extended over more than 2,600 km^2 (1,000 sq mi) and were home to about 800 rhinos. When poor farmers from the mid-hills moved to the Chitwan Valley in search of arable land, the area was subsequently opened for settlement, and poaching of wildlife became rampant. The Chitwan population has repeatedly been jeopardized by poaching; in 2002 alone, poachers killed 37 animals to saw off and sell their valuable horns.

Otters have been hunted for their pelts since at least the 1700s. There has been a long history of otter pelts being worn around the world. In China it was standard for the royalty to wear robes made from them. People that were financially high in status also wore them. Otters have also been hunted using dogs, specifically the otterhound. In modern times, TRAFFIC, a joint program of the World Wildlife Fund (WWF) and International Union for Conservation of Nature (IUCN), reported that otters are at serious risk in Southeast Asia and have disappeared from parts of their former range. This decline in populations is due to hunting to supply the demand for skins.

Habitat Degradation

The Baykalsk Pulp and Paper Mill was a major producer of industrial waste for Lake Baikal.

One problem at Lake Baikal is the introduction of pollutants into the ecosystem. Pesticides such as DDT and hexachlorocyclohexane, as well as industrial waste, mainly from the Baykalsk Pulp and Paper Mill, are thought to have been the cause of several disease epidemics among Baikal seal populations. The chemicals are speculated to concentrate up the food chain and weaken the Baikal seal's immune system, making them susceptible to diseases such as canine distemper and the plague, which was the cause of a serious Baikal seal epidemic that resulted in the deaths of thousands of animals in 1997 and 1999. Baikal seal pups have higher levels of DDT and PCB than known in any other population of European or Arctic earless seal.

In the 1940s, beavers were brought from Canada to the island of Tierra Del Fuego in southern Chile and Argentina, for commercial fur production. However, the project failed and the beavers, ten pairs, were released into the wild. Having no natural predators in their new environment, they quickly spread throughout the island, and to other islands in the region, reaching a number of 100,000 individuals within just 50 years. They are now considered a serious invasive species in the region, due to their massive destruction of forest trees, and efforts are being made for their eradication.

In some European countries, such as Belgium, France, and the Netherlands, the muskrat is considered an invasive pest, as its burrowing damages the dikes and levees on which these low-lying countries depend for protection from flooding. In those countries, it is trapped, poisoned, and hunted to attempt to keep the population down. Muskrats also eat corn and other farm and garden crops growing near water bodies.

Dead trees as a result of the construction of a beaver dam in Tierra del Fuego

Urban and agricultural development, increased damming, and increased use of hydroelectric power in rivers in countries such as Côte d'Ivoire and Ghana are threats to the African manatee's habitat and life, and thick congestion of boats in waterways may cause them to have a deadly run-in with the vessels. However, even natural occurrences, such as droughts and tidal changes, often strand manatees in an unsuitable habitat. Some are killed accidentally by fishing trawls and nets intended for catching sharks. The Amazonian manatee is at risk from pollution, accidental drowning in commercial fishing nets, and the degradation of vegetation by soil erosion resulting from deforestation. Additionally, the indiscriminate release of mercury in mining activities threatens the entire aquatic ecosystem of the Amazon Basin.

As China developed economically, pressure on the Chinese river dolphin grew significantly. Industrial and residential waste flowed into the Yangtze. The riverbed was dredged and reinforced with concrete in many locations. Ship traffic multiplied, boats grew in size, and fishermen employed wider and more lethal nets. Noise pollution caused the nearly blind animal to collide with propellers. Stocks of the dolphin's prey declined drastically in the late 20th century, with some fish populations declining to one thousandth of their pre-industrial levels. In the 1950s, the population was estimated at 6,000 animals, but declined rapidly over the subsequent five decades. Only a few hundred were left by 1970. Then the number dropped down to 400 by the 1980s and then to 13 in 1997 when a full-fledged search was conducted. On December 13, 2006, the baiji was declared functionally extinct, after a 45-day search by leading experts in the field failed to find a single specimen. The last verified sighting was in 2004.

As Food

Moose are hunted as a game species in many of the countries where they are found. While the flesh has protein levels similar to those of other comparable red meats (e.g. beef, deer and elk), it has a low fat content, and the fat that is present consists of a higher proportion of polyunsaturated fats rather than saturated fats.

> "like tender beef, with perhaps more flavour; sometimes like veal"
>
> —*Henry David Thoreau of The Maine Woods describing the taste of moose meat*

Cadmium levels are high in moose liver and kidneys, with the result that consumption of these organs from moose more than one year old is prohibited in Finland. Cadmium intake has been found to be elevated amongst all consumers of moose meat, though the meat was found to contribute only slightly to the daily cadmium intake. However the consumption of moose liver or kidneys significantly increased cadmium intake, with the study revealing that heavy consumers of moose organs have a relatively narrow safety margin below the levels which would probably cause adverse health effects.

In the 17th century, based on a question raised by the Bishop of Quebec, the Roman Catholic Church ruled that the beaver was a fish (beaver flesh was a part of the indigenous peoples' diet, prior to the Europeans' arrival) for purposes of dietary law. Therefore, the general prohibition on the consumption of meat on Fridays did not apply to beaver meat. This is similar to the Church's classification of other semi-aquatic rodents, such as the capybara and muskrat.

Marine Mammal

A Marine mammal is a mammal that has adapted to aquatic life and rely's on the ocean to maintain a healthy, livable existence.

These amazing animals can be found living in all of the worlds major oceans from the tropical environments in and around the equator to the northern and southern polar hemispheres in and around the Arctic and Antarctic oceans.

Marine mammals share several characteristics that are common among all mammals such as the need to breathe air, being warm-blooded, having mammary glands which produce milk to feed their young, giving birth to live young (pregnancy/gestation periods) and in some cases having hair.

Here are several common characteristics found in marine mammals:

- Marine mammals breathe air: Although marine mammals live in and around the water they must come to the surface to breathe otherwise they'll drown.
- Marine mammals are warm-blooded: In order to maintain their body heat marine mammals consume large quantities of calories and develop a thick layer of fat or blubber to keep their vital organs from freezing in cold environments.
- Marine mammals give birth to their offspring: Unlike fish and other aquatic animals whales do not lay eggs; instead they carry their children in their womb until they are born.
- Marine mammals produce milk: Marine mammals have mammary glands that produce milk which they use to feed their children. The milk is often full of fat and nutrients to help the child develop.

While land mammals are similar in many ways marine mammals have adapted to deal with living in environments that are generally surrounded by water rather than being elusively surrounded by land, and way marine mammals live, hunt for food and navigate the world is largely dependent upon their oceanic environment and physical characteristics.

For example marine mammals such as whales, dolphins and porpoises have streamlined bodies designed to reduce water resistance when swimming, they also have specialized lungs and muscles designed to store oxygen, thick blubber or fat to help maintain body heat in cold environments, large veins to help transfer blood to vital organs in a cold water or when in deep waters where compression can become an issue and fins, flukes, and flippers to assist with swimming.

Other species such as polar bears have thick layers of fat and fur to help them maintain their body heat in cold climates, a large body to help keep them warm and disperse their body heat, thick curved claws for gripping the ice and predators, sharp teeth and strong jaws for tearing apart prey and powerful muscles for fighting and attacking.

Seals have thick layers of fat/blubber to help keep them warm, rounded bodies to help disperse their body heat, limbs designed for walking/waddling on land with webbed feet/flippers to help them swim and teeth which they use to grip onto their prey.

Seals, sea lions and walruses have a number of characteristics that are similar among the separate species but also share a few distinct differences as well.

For example walruses tend to have a stockier build than seals and sea lions and they also possess two large teeth that protrude from their mouth, similar to a saber toothed tiger only larger.

Seals also have separate distinct characteristics from sea lions such as having shorter for limbs with long claws, which requires them to waddle on land rather than walk and their strong rare flippers allow them to propel themselves through the water more effectively than sea lions.

While not all marine mammals share the same biological characteristics, (a clear example would be to compare polar bears to whales) they all live near or in the ocean and need the ocean for its supply of food and water in order to survive.

The Distinct Types of Marine Mammals

In terms of the various marine mammal species there are several distinct groups of marine mammals; sirenians (manatees and dugongs) cetaceans (whales, dolphins and porpoises), fissipeds, sea otters, pinnipeds (seals, fur seals, sea lions and walruses) and (as considered by some researchers and biologists) polar bears.

Both cetaceans and sirenians live in the ocean while polar bears, pinnipeds and fissipeds are land dwellers, but rely on the ocean to supply them with food and water.

As a whole there are over 125 recorded species of marine mammal inhibiting the ocean and native aquatic environments of the world.

Here is a list of the of individual sub groups within the marine mammal family:

- Cetaceans (whales, dolphins and porpoises)
- Fissipeds
- Pinnipeds (seals, fur seals, sea lions and walruses)
- Ursidae (polar bears)
- Sea otters
- Sirenians (manatees and dugongs)

Although these animals can be found throughout the world the highest concentration of marine mammals (about 40%) are found at or around 40° both north and south of the equator.

Distribution and Habitat

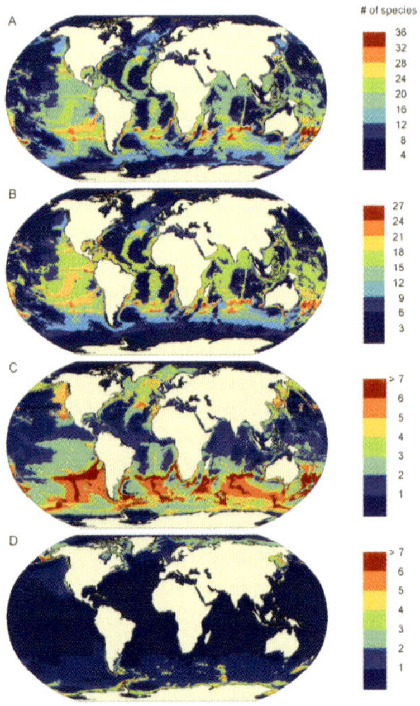

Marine mammal species richness: A) All species (n=115), B) Odontocetes (n=69), C) Mysticetes (n=14), D) Pinnipeds (n=32), based on data from 1990–1999

Marine mammals are widely distributed throughout the globe, but their distribution is patchy and coincides with the productivity of the oceans. Species richness peaks at around 40° latitude, both north and south. This corresponds to the highest levels of primary production around North and South America, Africa, Asia and Australia. Total species range is highly variable for marine mammal species. On average most marine mammals have ranges which are equivalent or smaller than one-fifth of the Indian Ocean. The variation observed in range size is a result of the different ecological requirements of each species and their ability to cope with a broad range of environmental conditions. The high degree of overlap between marine mammal species richness and areas of human impact on the environment is of concern.

Most marine mammals, such as seals and sea otters, inhabit the coast. Seals, however, also use a number of terrestrial habitats, both continental and island. In temperate and tropical areas, they haul-out on to sandy and pebble beaches, rocky shores, shoals, mud flats, tide pools and in sea caves. Some species also rest on man-made structures, like piers, jetties, buoys and oil platforms. Seals may move further inland and rest in sand dunes or vegetation, and may even climb cliffs. Most cetaceans live in the open ocean, and species like the sperm whale may dive to depths of −1,000 to −2,500 feet (−300 to −760 m) in search of food. Sirenians live in shallow coastal waters, usually living 30 feet (9.1 m) below sea level. However, they have been known to dive to −120 feet (−37 m) to forage deep-water seagrasses. Sea otters live in protected areas, such as rocky shores, kelp forests, and barrier reefs, although they may reside among drift ice or in sandy, muddy, or silty areas.

Many marine mammals seasonally migrate. Annual ice contains areas of water that appear and disappear throughout the year as the weather changes, and seals migrate in response to these changes. In turn, polar bears must follow their prey. In Hudson Bay, James Bay, and some other areas, the ice melts completely each summer (an event often referred to as "ice-floe breakup"), forcing polar bears to go onto land and wait through the months until the next freeze-up. In the Chukchi and Beaufort seas, polar bears retreat each summer to the ice further north that remains frozen year-round. Seals may also migrate to other environmental changes, such as El Niño, and traveling seals may use various features of their environment to reach their destination including geomagnetic fields, water and wind currents, the position of the sun and moon and the taste and temperature of the water. Baleen whales famously migrate very long distances into tropical waters to give birth and raise young, possibly to prevent predation by killer whales. The gray whale has the longest recorded migration of any mammal, with one traveling 14,000 miles (23,000 km) from the Sea of Okhotsk to the Baja Peninsula. During the winter, manatees living at the northern end of their range migrate to warmer waters.

Adaptations

Marine mammals have a number of physiological and anatomical features to overcome the unique challenges associated with aquatic living. Some of these features are very species specific. Marine mammals have developed a number of features for efficient locomotion such as torpedo shaped bodies to reduce drag; modified limbs for propulsion and steering; tail flukes and dorsal fins for propulsion and balance. Marine mammals are adept at thermoregulation using dense fur or blubber, circulatory adjustments (counter-current heat exchangers); torpedo shaped bodies, reduced appendages, and large size to prevent heat loss.

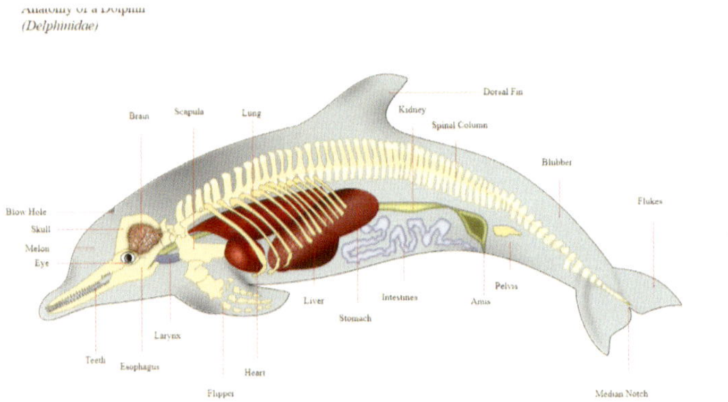

The anatomy of a dolphin showing its skeleton, major organs, and body shape

Marine mammals are able to dive for long periods of time. Both pinnipeds and cetaceans have large and complex blood vessel systems which serve to store oxygen to support deep diving. Other important reservoirs include muscles, blood, and the spleen which all have the capacity to hold a high concentration of oxygen. They are also capable of bradycardia (reduced heart rate), and vasoconstriction (shunting most of the oxygen to vital organs such as the brain and heart) to allow extended diving times and cope with oxygen deprivation. If oxygen is depleted, marine mammals can access substantial reservoirs of glycogen that support anaerobic glycolysis of the cells involved during conditions of systemic hypoxia associated with prolonged submersion.

Sound travels differently through water, and therefore marine mammals have developed adaptations to ensure effective communication, prey capture, and predator detection. The most notable adaptation is the development of echolocation in whales and dolphins. Toothed whales emit a focused beam of high-frequency clicks in the direction that their head is pointing. Sounds are generated by passing air from the bony nares through the phonic lips. These sounds are reflected by the dense concave bone of the cranium and an air sac at its base. The focused beam is modulated by a large fatty organ known as the 'melon'. This acts like an acoustic lens because it is composed of lipids of differing densities.

Marine mammals have evolved a wide variety of features for feeding, which are mainly seen in their dentition. For example, the cheek teeth of pinnipeds and odontocetes are specifically adapted to capture fish and squid. In contrast, baleen whales have evolved baleen plates to filter feed plankton and small fish from the water.

Polar bears, otters, and fur seals have fur, one of the defining mammalian features, that is long, oily, and waterproof in order to trap air to provide insulation. In contrast, other marine mammals – such as whales, dolphins, porpoises, manatees, dugongs, and walruses – have lost long fur in favor of a thick, dense epidermis and a thickened fat layer (blubber) in response to hydrodynamic requirements. Wading and bottom-feeding animals (such as manatees) need to be heavier than water in order to keep contact with the floor or to stay submerged. Surface-living animals (such as sea otters) need the opposite, and free-swimming animals living in open waters (such as dolphins) need to be neutrally buoyant in order to be able to swim up and down the water column. Typically, thick and dense bone is found in bottom feeders and low bone density is associated with mammals living in deep water. Some marine mammals, such as polar bears and otters, have retained four weight-bearing limbs and can walk on land like fully terrestrial animals.

Ecology

Diet

Killer whale hunting a Weddel seal

All cetaceans are carnivorous and predatory. Toothed whales mostly feed on fish and cephalopods, followed by crustaceans and bivalves. Some may forage with other kinds of animals, such as other species of whales or certain species of pinnipeds. One common feeding method is herding, where a pod squeezes a school of fish into a small volume, known as a bait ball. Individual members then take turns plowing through the ball, feeding on the stunned fish. Coralling is a method where dolphins chase fish into shallow water to catch them more easily. Killer whales and bottlenose dolphins have also been known to drive their prey onto a beach to feed on it. Other whales with a blunt snout and reduced dentition rely on suction feeding. Though carnivorous, they house gut flora similar to that of terrestrial herbivores, probably a remnant of their herbivorous ancestry.

Baleen whales use their baleen plates to sieve plankton, among others, out of the water; there are two types of methods: lunge-feeding and gulp-feeding. Lunge-feeders expand the volume of their jaw to a volume bigger than the original volume of the whale itself by inflating their mouth. This causes grooves on their throat to expand, increasing the amount of water the mouth can store. They ram a bait ball at high speeds in order to feed, but this is only energy-effective when used against a large bait ball. Gulp-feeders swim with an open mouth, filling it with water and prey. Prey must occur in sufficient numbers to trigger the whale's interest, be within a certain size range so that the baleen plates can filter it, and be slow enough so that it cannot escape.

Sea otters have dexterous hands which they use to smash sea urchins off rocks.

Otters are the only marine animals that are capable of lifting and turning over rocks, which they often do with their front paws when searching for prey. The sea otter may pluck snails and other organisms from kelp and dig deep into underwater mud for clams. It is the only marine mammal that catches fish with its forepaws rather than with its teeth. Under each foreleg, sea otters have a loose pouch of skin that extends across the chest which they use to store collected food to bring to the surface. This pouch also holds a rock that is used to break open shellfish and clams, an example of tool use. The sea otters eat while floating on their backs, using their forepaws to tear food apart and bring to their mouths. Marine otters mainly feed on crustaceans and fish.

Pinnipeds mostly feed on fish and cephalopods, followed by crustaceans and bivalves, and then zooplankton and warm-blooded prey (like sea birds). Most species are generalist feeders, but a few are specialists. They typically hunt non-schooling fish, slow-moving or immobile invertebrates or endothermic prey when in groups. Solitary foraging species usually exploit coastal waters, bays and rivers. When large schools of fish or squid are available, pinnipeds hunt cooperatively in large groups, locating and herding their prey. Some species, such as California and South American sea lions, may forage with cetaceans and sea birds.

The polar bear is the most carnivorous species of bear, and its diet primarily consists of ringed (*Pusa hispida*) and bearded (*Erignathus barbatus*) seals. Polar bears hunt primarily at the interface between ice, water, and air; they only rarely catch seals on land or in open water. The polar bear's most common hunting method is still-hunting: The bear locates a seal breathing hole using its sense of smell, and crouches nearby for a seal to appear. When the seal exhales, the bear smells its breath, reaches into the hole with a forepaw, and drags it out onto the ice. The polar bear also hunts by stalking seals resting on the ice. Upon spotting a seal, it walks to within 100 yards (90 m), and then crouches. If the seal does not notice, the bear creeps to within 30 to 40 feet (9 to 10 m) of the seal and then suddenly rushes to attack. A third hunting method is to raid the birth lairs that female seals create in the snow. They may also feed on fish.

A dugong feeding on the sea-floor

Sirenians are referred to as "sea cows" because their diet consists mainly of sea-grass. When eating, they ingest the whole plant, including the roots, although when this is impossible they feed on just the leaves. A wide variety of seagrass has been found in dugong stomach contents, and evidence exists they will eat algae when seagrass is scarce. West Indian manatees eat up to 60 different species of plants, as well as fish and small invertebrates to a lesser extent.

Keystone Species

Sea otters are a classic example of a keystone species; their presence affects the ecosystem more profoundly than their size and numbers would suggest. They keep the population of certain benthic (sea floor) herbivores, particularly sea urchins, in check. Sea urchins graze on the lower stems of kelp, causing the kelp to drift away and die. Loss of the habitat and nutrients provided by kelp forests leads to profound cascade effects on the marine ecosystem. North Pacific areas that do not have sea otters often turn into urchin barrens, with abundant sea urchins and no kelp forest. Reintroduction of sea otters to British Columbia has led to a dramatic improvement in the health of coastal ecosystems, and similar changes have been observed as sea otter populations recovered in the Aleutian and Commander Islands and the Big Sur coast of California However, some kelp forest ecosystems in California have also thrived without sea otters, with sea urchin populations apparently controlled by other factors. The role of sea otters in maintaining kelp forests has been observed to be more important in areas of open coast than in more protected bays and estuaries.

Antarctic fur seal pups (left) vs. Arctic harp seal pup (right)

An apex predator affects prey population dynamics and defense tactics (such as camouflage). The polar bear is the apex predator within its range. Several animal species, particularly Arctic foxes (*Vulpes lagopus*) and glaucous gulls (*Larus hyperboreus*), routinely scavenge polar bear kills. The relationship between ringed seals and polar bears is so close that the abundance of ringed seals in some areas appears to regulate the density of polar bears, while polar bear predation in turn regulates density and reproductive success of ringed seals. The evolutionary pressure of polar bear predation on seals probably accounts for some significant differences between Arctic and Antarctic seals. Compared to the Antarctic, where there is no major surface predator, Arctic seals use more breathing holes per individual, appear more restless when hauled out on the ice, and rarely defecate on the ice. The fur of Arctic pups is white, presumably to provide camouflage from predators, whereas Antarctic pups all have dark fur.

Killer whales are apex predators throughout their global distribution, and can have a profound effect on the behavior and population of prey species. Their diet is very broad and they can feed on many vertebrates in the ocean including salmon, rays, sharks (even white sharks), large baleen whales, and nearly 20 species of pinniped. The predation of whale calves may be responsible for annual whale migrations to calving grounds in more tropical waters, where the population of killer whales is much lower than in polar waters. Prior to whaling, it is thought that great whales were a major food source; however, after their sharp decline, killer whales have since expanded their diet, leading to the decline of smaller marine mammals. A decline in Aleutian Islands sea otter populations in the 1990s was controversially attributed by some scientists to killer whale predation, although with no direct evidence. The decline of sea otters followed a decline in harbor seal and

Steller sea lion populations, the killer whale's preferred prey, which in turn may be substitutes for their original prey, now reduced by industrial whaling.

Whale Pump

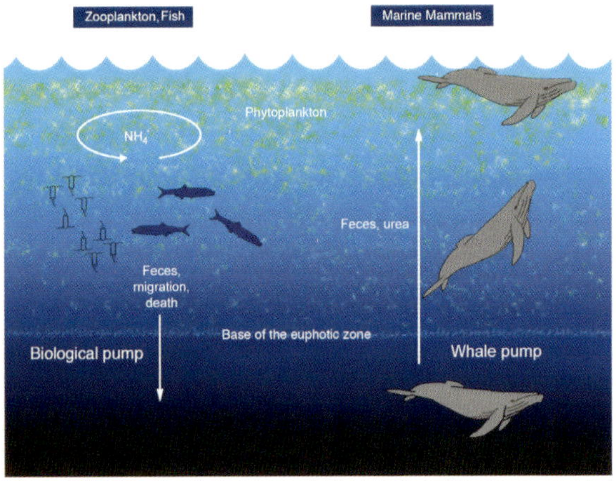

"Whale pump" – the role played by whales in recycling ocean nutrients

A 2010 study considered whales to be a positive influence to the productivity of ocean fisheries, in what has been termed a "whale pump". Whales carry nutrients such as nitrogen from the depths back to the surface. This functions as an upward biological pump, reversing an earlier presumption that whales accelerate the loss of nutrients to the bottom. This nitrogen input in the Gulf of Maine is more than the input of all rivers combined emptying into the gulf, some 25,000 short tons (23,000 t) each year. Whales defecate at the ocean's surface; their excrement is important for fisheries because it is rich in iron and nitrogen. The whale feces are liquid and instead of sinking, they stay at the surface where phytoplankton feed off it.

Upon death, whale carcasses fall to the deep ocean and provide a substantial habitat for marine life. Evidence of whale falls in present-day and fossil records shows that deep sea whale falls support a rich assemblage of creatures, with a global diversity of 407 species, comparable to other neritic biodiversity hotspots, such as cold seeps and hydrothermal vents. Deterioration of whale carcasses happens though a series of three stages. Initially, moving organisms, such as sharks and hagfish, scavenge soft tissue at a rapid rate over a period of months to as long as two years. This is followed by the colonization of bones and surrounding sediments (which contain organic matter) by enrichment opportunists, such as crustaceans and polychaetes, throughout a period of years. Finally, sulfophilic bacteria reduce the bones releasing hydrogen sulphide enabling the growth of chemoautotrophic organisms, which in turn, support other organisms such as mussels, clams, limpets, and sea snails. This stage may last for decades and supports a rich assemblage of species, averaging 185 species per site.

Interactions with Humans

Exploitation

Marine mammals were hunted by coastal aboriginal humans historically for food and other resources. These subsistence hunts still occur in Canada, Greenland, Indonesia, Russia, the United

States, and several nations in the Caribbean. The effects of these are only localized, as hunting efforts were on a relatively small scale. Commercial hunting took this to a much greater scale and marine mammals were heavily exploited. This led to the extinction of the †Steller's sea cow (*Hydrodamalis gigas*), †sea mink (*Neovison macrodon*), †Japanese sea lion (*Zalophus japonicus*), and the †Caribbean monk seal (*Neomonachus tropicalis*). Today, populations of species that were historically hunted, such as blue whales (*Balaenoptera musculus*) and the North Pacific right whale (*Eubalaena japonica*), are much lower than their pre-whaling levels. Because whales generally have slow growth rates, are slow to reach sexual maturity, and have a low reproductive output, population recovery has been very slow.

Men killing northern fur seals on Saint Paul Island, Alaska in the 1890s

A number of whales are still subject to direct hunting, despite the 1986 moratorium ban on whaling set under the terms of the International Whaling Commission (IWC). There are only two nations remaining which sanction commercial whaling: Norway, where several hundred common minke whales are harvested each year; and Iceland, where quotas of 150 fin whales and 100 minke whales per year are set. Japan also harvests several hundred Antarctic and North Pacific minke whales each year, ostensibly for scientific research in accordance with the moratorium. However, the illegal trade of whale and dolphin meat is a significant market in Japan and some countries.

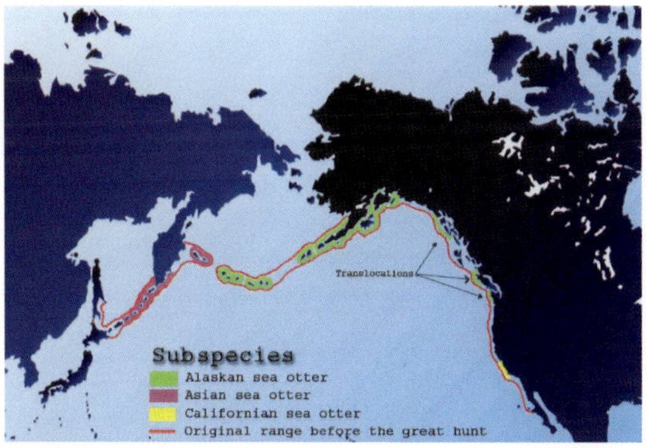

Historical and modern range of northern sea otters

The most profitable furs in the fur trade were those of sea otters, especially the northern sea otter which inhabited the coastal waters between the Columbia River to the south and Cook Inlet to the north. The fur of the Californian southern sea otter was less highly prized and thus less profitable. After the northern sea otter was hunted to local extinction, maritime fur traders shifted to California until the southern sea otter was likewise nearly extinct. The British and American maritime fur traders took their furs to the Chinese port of Guangzhou (Canton), where they worked within the established Canton System. Furs from Russian America were mostly sold to China via the

Mongolian trading town of Kyakhta, which had been opened to Russian trade by the 1727 Treaty of Kyakhta.

Commercial sealing was historically just as important as the whaling industry. Exploited species included harp seals, hooded seals, Caspian seals, elephant seals, walruses and all species of fur seal. The scale of seal harvesting decreased substantially after the 1960s, after the Canadian government reduced the length of the hunting season and implemented measures to protect adult females. Several species that were commercially exploited have rebounded in numbers; for example, Antarctic fur seals may be as numerous as they were prior to harvesting. The northern elephant seal was hunted to near extinction in the late 19th century, with only a small population remaining on Guadalupe Island. It has since recolonized much of its historic range, but has a population bottleneck. Conversely, the Mediterranean monk seal was extirpated from much of its former range, which stretched from the Mediterranean to the Black Sea and northwest Africa, and only remains in the northeastern Mediterranean and some parts of northwest Africa.

Polar bears can be hunted for sport in Canada with a special permit and accompaniment by a local guide. This can be an important source of income for small communities, as guided hunts bring in more income than selling the polar bear hide on markets. The United States, Russia, Norway, Greenland, and Canada allow subsistence hunting, and Canada distributes hunting permits to indigenous communities. The selling of these permits is a main source of income for many of these communities. Their hides can be used for subsistence purposes, kept as hunting trophies, or can be bought in markets.

Ocean Traffic and Fisheries

The remains of a North Atlantic right whale after it collided with a ship propeller

By-catch is the incidental capture of non-target species in fisheries. Fixed and drift gill nets cause the highest mortality levels for both cetaceans and pinnipeds, however, entanglements in long lines, mid-water trawls, and both trap and pot lines are also common. Tuna seines are particularly problematic for entanglement by dolphins. By-catch affects all cetaceans, both small and big, in all habitat types. However, smaller cetaceans and pinnipeds are most vulnerable as their size means that escape once they are entangled is highly unlikely and they frequently drown. While larger cetaceans are capable of dragging nets with them, the nets sometimes remain tightly attached to the individual and can impede the animal from feeding sometimes leading to starvation. Abandoned or lost nets and lines cause mortality through ingestion or entanglement. Marine mammals also get entangled in aquaculture nets, however, these are rare events and not prevalent enough to impact populations.

Vessel strikes cause death for a number of marine mammals, especially whales. In particular, fast commercial vessels such as container ships can cause major injuries or death when they collide with marine mammals. Collisions occur both with large commercial vessels and recreational boats and cause injury to whales or smaller cetaceans. The critically endangered North Atlantic right whale is particularly affected by vessel strikes. Tourism boats designed for whale and dolphin watching can also negatively impact on marine mammals by interfering with their natural behavior.

The fishery industry not only threatens marine mammals through by-catch, but also through competition for food. Large scale fisheries have led to the depletion of fish stocks that are important prey species for marine mammals. Pinnipeds have been especially affected by the direct loss of food supplies and in some cases the harvesting of fish has led to food shortages or dietary deficiencies, starvation of young, and reduced recruitment into the population. As the fish stocks have been depleted, the competition between marine mammals and fisheries has sometimes led to conflict. Large-scale culling of populations of marine mammals by commercial fishers has been initiated in a number of areas in order to protect fish stocks for human consumption.

Shellfish aquaculture takes up space so in effect creates competition for space. However, there is little direct competition for aquaculture shellfish harvest. On the other hand, marine mammals regularly take finfish from farms, which creates significant problems for marine farmers. While there are usually legal mechanisms designed to deter marine mammals, such as anti-predator nets or harassment devices, individuals are often illegally shot.

Habitat Loss and Degradation

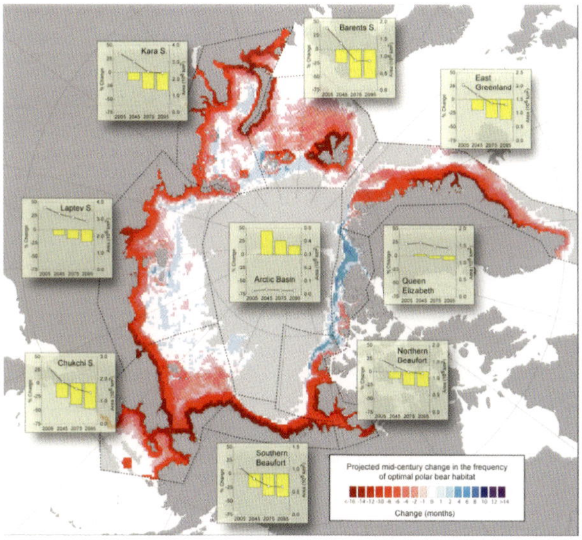

Map from the U.S. Geological Survey shows projected changes in polar bear habitat from 2005 to 2095. Red areas indicate loss of optimal polar bear habitat; blue areas indicate gain.

Habitat degradation is caused by a number of human activities. Marine mammals that live in coastal environments are most likely to be affected by habitat degradation and loss. Developments such as sewage marine outfalls, moorings, dredging, blasting, dumping, port construction, hydroelectric projects, and aquaculture both degrade the environment and take up valuable habitat. For example, extensive shellfish aquaculture takes up valuable space used by coastal marine mammals for important activities such as breeding, foraging and resting.

Contaminants that are discharged into the marine environment accumulate in the bodies of marine mammals when they are stored unintentionally in their blubber along with energy. Contaminants that are found in the tissues of marine mammals include heavy metals, such as mercury and lead, but also organochlorides and polycyclic aromatic hydrocarbons. For example, these can cause disruptive effects on endocrine systems; impair the reproductive system, and lower the immune system of individuals, leading to a higher number of deaths. Other pollutants such as oil, plastic debris and sewage threaten the livelihood of marine mammals.

Noise pollution from anthropogenic activities is another major concern for marine mammals. This is a problem because underwater noise pollution interferes with the abilities of some marine mammals to communicate, and locate both predators and prey. Underwater explosions are used for a variety of purposes including military activities, construction and oceanographic or geophysical research. They can cause injuries such as hemorrhaging of the lungs, and contusion and ulceration of the gastrointestinal tract. Underwater noise is generated from shipping, the oil and gas industry, research, and military use of sonar and oceanographic acoustic experimentation. Acoustic harassment devices and acoustic deterrent devices used by aquaculture facilities to scare away marine mammals emit loud and noxious underwater sounds.

Two changes to the global atmosphere due to anthropogenic activity threaten marine mammals. The first is increases in ultraviolet radiation due to ozone depletion, and this mainly affects the Antarctic and other areas of the southern hemisphere. An increase in ultraviolet radiation has the capacity to decrease phytoplankton abundance, which forms the basis of the food chain in the ocean. The second effect of global climate change is global warming due to increased carbon dioxide levels in the atmosphere. Raised sea levels, sea temperature and changed currents are expected to affect marine mammals by altering the distribution of important prey species, and changing the suitability of breeding sites and migratory routes. The Arctic food chain would be disrupted by the near extinction or migration of polar bears. Arctic sea ice is the polar bear's habitat. It has been declining at a rate of 13% per decade because the temperature is rising at twice the rate of the rest of the world. By the year 2050, up to two-thirds of the world's polar bears may vanish if the sea ice continues to melt at its current rate.

A study by evolutionary biologists at the University of Pittsburgh showed that the ancestors of many marine mammals stopped producing a certain enzyme that today protects against some neurotoxic chemicals called organophosphates,, including those found in the widely used pesticides chlorpyrifos and diazinon. Marine mammals may be increasingly exposed to these compounds due to agricultural runoff reaching the world's oceans.

Protection

The Marine Mammal Protection Act of 1972 (MMPA) was passed on October 21, 1972 under president Richard Nixon to prevent the further depletion and possible extinction of marine mammal stocks. It prohibits the taking ("the act of hunting, killing, capture, and/or harassment of any marine mammal; or, the attempt at such") of any marine mammal without a permit issued by the Secretary. Authority to manage the MMPA was divided between the Secretary of the Interior through the U.S. Fish and Wildlife Service (Service), and the Secretary of Commerce, which is delegated to the National Oceanic and Atmospheric Administration (NOAA). The Marine Mammal Commission (MMC) was established to review existing policies and make recommendations to the Service and NOAA to better implement the MMPA. The Service is responsible for ensuring the protection of sea otters and marine otters, walruses, polar bears, the three species of manatees, and dugongs; and NOAA was given responsibility to conserve and manage pinnipeds (excluding walruses) and cetaceans.

Signatory countries of the International Whaling Commission (IWC)

The Act was updated on 1 January 2016 with a clause banning “the import of fish from fisheries that cannot prove they meet US standards for protecting marine mammals.” The requirement to show that protection standards are met is hoped to compel countries exporting fish to the US to more strictly control their fisheries that no protected marine mammals are adversely affected by fishing.

The 1979 Convention on the Conservation of Migratory Species of Wild Animals (CMS) is the only global organization that conserves a broad range of animals, of which includes marine mammals. Of the agreements made, three of them deal with the conservation of marine mammals: ACCOBAMS, ASCOBANS, and the Wadden Sea Agreement. In 1982, the United Nations Convention on the Law of the Sea (LOSC) adopted a pollution prevention approach to conservation, which many other conventions at the time also adopted.

The Agreement on the Conservation of Cetaceans in the Black Sea, Mediterranean Sea and contiguous Atlantic area (ACCOBAMS), founded in 1996, specifically protects cetaceans in the Mediterranean area, and “maintains a favorable status”, a direct action against whaling. There are 23 member states. The Agreement on the Conservation of Small Cetaceans of the Baltic and North Seas (ASCOBANS) was adopted alongside ACCOBAMS to establish a special protection area for Europe’s increasingly threatened cetaceans. Other anti-whaling efforts include a ten-year moratorium in 1986 by the IWC on all whaling, and an environmental agreement (a type of international law) the International Convention for the Regulation of Whaling which controlled commercial, scientific, and subsistence whaling.

The Agreement on the Conservation of Seals in the Wadden Sea, enforced in 1991, prohibits the killing or harassment of seals in the Wadden Sea, specifically targeting the harbor seal population.

The 1973 Agreement on the Conservation of Polar Bears between Canada, Denmark (Greenland), Norway (Svalbard), the United States, and the Soviet Union outlawed the unregulated hunting of polar bears from aircraft and icebreakers, as well as protecting migration, feeding, and hibernation sites.

Various non-governmental organizations participate in marine conservation activism, wherein they draw attention to and aid in various problems in marine conservation, such as pollution, whaling, bycatch, and so forth. Notable organizations include the Greenpeace who focus on overfishing and whaling among other things, and Sea Shepherd Conservation Society who are known for taking direct-action tactics to expose illegal activity.

As Food

Pilot whale meat (bottom), blubber (middle) and dried fish (left) with potatoes, Faroe Islands

For thousands of years, indigenous peoples of the Arctic have depended on whale meat. The meat is harvested from legal, non-commercial hunts that occur twice a year in the spring and autumn. The meat is stored and eaten throughout the winter. The skin and blubber (muktuk) taken from the bowhead, beluga, or narwhal is also valued, and is eaten raw or cooked. Whaling has also been practiced in the Faroe Islands in the North Atlantic since about the time of the first Norse settlements on the islands. Around 1000 long-finned pilot whales are still killed annually, mainly during the summer. Today, dolphin meat is consumed in a small number of countries worldwide, which include Japan and Peru (where it is referred to as *chancho marino*, or "sea pork"). In some parts of the world, such as Taiji, Japan and the Faroe Islands, dolphins are traditionally considered food, and are killed in harpoon or drive hunts.

There have been human health concerns associated with the consumption of dolphin meat in Japan after tests showed that dolphin meat contained high levels of methyl mercury. There are no known cases of mercury poisoning as a result of consuming dolphin meat, though the government continues to monitor people in areas where dolphin meat consumption is high. The Japanese government recommends that children and pregnant women avoid eating dolphin meat on a regular basis. Similar concerns exist with the consumption of dolphin meat in the Faroe Islands, where prenatal exposure to methyl mercury and PCBs primarily from the consumption of pilot whale meat has resulted in neuropsychological deficits amongst children.

> The Faroe Islands population was exposed to methyl mercury largely from contaminated pilot whale meat, which contained very high levels of about 2 mg methyl mercury/kg. However, the Faroe Islands populations also eat significant amounts of fish. The study of about 900 Faroese children showed that prenatal exposure to methyl mercury resulted in neuropsychological deficits at 7 years of age
>
> — *World Health Organization*

Ringed seals were once the main food staple for the Inuit. They are still an important food source for the people of Nunavut and are also hunted and eaten in Alaska. Seal meat is an important source of food for residents of small coastal communities. The seal blubber is used to make seal oil, which is marketed as a fish oil supplement. In 2001, two percent of Canada's raw seal oil was processed and sold in Canadian health stores.

In Captivity

Aquariums

- Cetaceans

Performing killer whale at SeaWorld San Diego, 2009

Various species of dolphins are kept in captivity. These small cetaceans are more often than not kept in theme parks and dolphinariums, such as SeaWorld. Bottlenose dolphins are the most common species of dolphin kept in dolphinariums as they are relatively easy to train and have a long lifespan in captivity. Hundreds of bottlenose dolphins live in captivity across the world, though exact numbers are hard to determine. The dolphin "smile" makes them popular attractions, as this is a welcoming facial expression in humans; however the smile is due to a lack of facial muscles and subsequent lack of facial expressions.

Organizations such as World Animal Protection and the Whale and Dolphin Conservation Society campaign against the practice of keeping cetaceans, particularly killer whales, in captivity. In captivity, they often develop pathologies, such as the dorsal fin collapse seen in 60–90% of male killer whales. Captives have vastly reduced life expectancies, on average only living into their 20s. In the wild, females who survive infancy live 46 years on average, and up to 70–80 years in rare cases. Wild males who survive infancy live 31 years on average, and up to 50–60 years. Captivity usually bears little resemblance to wild habitat, and captive whales' social groups are foreign to those found in the wild. Captive life is also stressful due the requirement to perform circus tricks that are not part of wild killer whale behavior, as well as restricting pool size. Wild killer whales may travel up to 100 miles (160 km) in a day, and critics say the animals are too big and intelligent to be suitable for captivity. Captives occasionally act aggressively towards themselves, their tankmates, or humans, which critics say is a result of stress. Dolphins are often trained to do several anthropomorphic behaviors, including waving and kissing—behaviors wild dolphins would rarely do.

- Pinnipeds

A sea lion trained to balance a ball on its nose

The large size and playfulness of pinnipeds make them popular attractions. Some exhibits have rocky backgrounds with artificial haul-out sites and a pool, while others have pens with small rocky, elevated shelters where the animals can dive into their pools. More elaborate exhibits contain deep pools that can be viewed underwater with rock-mimicking cement as haul-out areas. The most common pinniped species kept in captivity is the California sea lion as it is abundant and easy to train. These animals are used to perform tricks and entertain visitors. Other species popularly kept in captivity include the grey seal and harbor seal. Larger animals like walruses and Steller sea lions are much less common. Pinnipeds are popular attractions because they are "disneyfied", and, consequently, people often anthropomorphize them with a curious, funny, or playful nature.

Some organizations, such as the Humane Society of the United States and World Animal Protection, object to keeping pinnipeds and other marine mammals in captivity. They state that the exhibits could not be large enough to house animals that have evolved to be migratory, and a pool could never replace the size and biodiversity of the ocean. They also oppose using sea lions for entertainment, claiming the tricks performed are "exaggerated variations of their natural behaviors" and distract the audience from the animal's unnatural environment.

- Otter

Sea otters can do well in captivity, and are featured in over 40 public aquariums and zoos. The Seattle Aquarium became the first institution to raise sea otters from conception to adulthood with the birth of Tichuk in 1979, followed by three more pups in the early 1980s. In 2007, a YouTube video of two cute sea otters holding paws drew 1.5 million viewers in two weeks, and had over 20 million views as of January 2015. Filmed five years previously at the Vancouver Aquarium, it was YouTube's most popular animal video at the time, although it has since been surpassed. Otters are often viewed as having a "happy family life", but this is an anthropomorphism.

- Sirenians

The oldest manatee in captivity was Snooty, at the South Florida Museum's Parker Manatee Aquarium in Bradenton, Florida. Born at the Miami Aquarium and Tackle Company on July 21, 1948,

Snooty was one of the first recorded captive manatee births. He was raised entirely in captivity, and died at the age of 69. Manatees can also be viewed in a number of European zoos, such as the Tierpark in Berlin, the Nuremberg Zoo, in ZooParc de Beauval in France, and in the Aquarium of Genoa in Italy. The River Safari at Singapore features seven of them.

Military

A dolphin wearing a locating pinger, performing mine clearance work in the Iraq War

Bottlenose dolphins and California sea lions were used in the United States Navy Marine Mammal Program (NMMP) to detect mines, protect ships from enemy soldiers, and recover objects. The Navy has never trained attack dolphins, as they would not be able to discern allied soldiers from enemy soldiers. There were five marine mammal teams, each purposed for one of the three tasks: MK4 (dolphins), MK5 (sea lions), MK6 (dolphins and sea lions), MK7 (dolphins), and MK8 (dolphins); MK is short for mark. The dolphin teams were trained to detect and mark mines either attached to the seafloor or floating in the water column, because dolphins can use their echolocative abilities to detect mines. The sea lion team retrieved test equipment such as fake mines or bombs dropped from planes usually out of reach of divers who would have to make multiple dives. MK6 protects harbors and ships from enemy divers, and was operational in the Gulf War and Vietnam War. The dolphins would swim up behind enemy divers and attach a buoy to their air tank, so that they would float to the surface and alert nearby Navy personnel. Sea lions would hand-cuff the enemy, and try to outmaneuver their counter-attacks.

The use of marine mammals by the Navy, even in accordance with the Navy's policy, continues to meet opposition. The Navy's policy says that only positive reinforcement is to be used while training the military dolphins, and that they be cared for in accordance with accepted standards in animal care. The inevitable stresses involved in training are topics of controversy, as their treatment is unlike the animals' natural lifestyle, especially towards their confined spaces when not training. There is also controversy over the use of muzzles and other inhibitors, which prevent the dolphins from foraging for food while working. The Navy states that this is to prevent them from ingesting harmful objects, but conservation activists say this is done to reinforce the trainers' control over the dolphins, who hand out food rewards. The means of transportation is also an issue for conservation activists, since they are hauled in dry carriers, and switching tanks and introducing the dolphin to new dolphins is potentially dangerous as they are territorial.

Blue Whale

Blue whale (Balaenoptera musculus) is also called sulfur-bottom whale which is the most massive animal ever to have lived within a species of baleen whale that weighs approximately 150 tons and may attain a length of more than 30 meters (98 feet). The largest accurately measured blue whale was a 29.5-metre female that weighed 180 metric tons (nearly 200 short [U.S.] tons), but there are reports of 33-metre catches that may have reached 200 metric tons. The heart of one blue whale was recorded at nearly 700 kg (about 1,500 pounds).

Blue whales are blue-gray in color with lighter gray mottling in the form of large spots, which appear as if they were dabbed on with a huge paintbrush. The lower surfaces of the flippers are lighter gray or white in some instances. The blue whale has been called the sulfur-bottom whale because of the yellowish underside of some individuals that is reminiscent of the pale yellow color of that chemical element; this coloration is imparted by certain algae (diatoms) living on the whale's body. The blue whale has a wide head, a small dorsal fin located near the fluke, and 80–100 long grooves running lengthwise down the throat and chest. Its mouth contains up to 800 plates of short, wide, black baleen, or "whalebone," with thick, coarse bristles used for catching food. Females are generally larger than males, and the largest animals live in the Southern Ocean around Antarctica.

A blue whale surfacing in the ocean

The blue whale is found alone or in small groups in all oceans, but populations in the Southern Hemisphere are much larger. In the Northern Hemisphere, blue whales can be seen regularly in the Gulf of St. Lawrence and off the coasts of Monterey, California, and Baja California, Mexico. They spend the summer in polar waters, feeding on shrimplike crustaceans called krill. During a dive, the blue whale may engage in a series of turns and 360° rolls to locate prey and rapidly reorient its body to sweep up large concentrations of krill in a single open-mouthed lunge. A single adult blue may consume as much as eight tons of krill per day. In the winter blue whales move toward the Equator to breed. After a gestation of about 12 months, one calf about 8 meters (about 26 feet) long is born in temperate waters. While nursing, calves gain up to 90 kg (about 198 pounds) per day on the rich milk of their mothers. Young are weaned after seven to eight months, when they have reached a length of about 15 meters (about 49 feet).

Behavior

Feeding

Blue whales feed almost exclusively on krill, though they also take small numbers of copepods. The species of this zooplankton eaten by blue whales varies from ocean to ocean. In the North Atlantic, *Meganyctiphanes norvegica, Thysanoessa raschii, Thysanoessa inermis* and *Thysanoessa longicaudata* are the usual food; in the North Pacific, *Euphausia pacifica, Thysanoessa inermis, Thysanoessa longipes, Thysanoessa spinifera, Nyctiphanes symplex* and *Nematoscelis megalops*; and in the Southern Hemisphere, *Euphausia superba, Euphausia crystallorophias, Euphausia valentini*, and *Nyctiphanes australis*.

An adult blue whale can eat up to 40 million krill in a day. The whales always feed in the areas with the highest concentration of krill, sometimes eating up to 3,600 kilograms (7,900 lb) of krill in a single day. The daily energy requirement of an adult blue whale is in the region of 1.5 million kilocalories (6.3 GJ). Their feeding habits are seasonal. Blue whales gorge on krill in the rich waters of the Antarctic before migrating to their breeding grounds in the warmer, less-rich waters nearer the equator. The blue whale can take in up to 90 times as much energy as it expends, allowing it to build up considerable energy reserves.

Because krill move, blue whales typically feed at depths of more than 100 metres (330 ft) during the day and only surface-feed at night. Dive times are typically 10 minutes when feeding, though dives of up to 21 minutes are possible. The whale feeds by lunging forward at groups of krill, taking the animals and a large quantity of water into its mouth. The water is then squeezed out through the baleen plates by pressure from the ventral pouch and tongue. Once the mouth is clear of water, the remaining krill, unable to pass through the plates, are swallowed. The blue whale also incidentally consumes small fish, crustaceans and squid caught up with krill.

Life History

A blue whale calf with its mother

Mating starts in late autumn and continues to the end of winter. Little is known about mating behavior or breeding grounds. In the fall, males will follow females for prolonged periods of time. Occasionally, a second male will attempt to displace the first, and the whales will race each other at high speed, ranging from 17 miles per hour (27 km/h) to 20 miles per hour (32 km/h) in New Zealand. This often causes the racing whales to breach, which is rare in blue whales. This racing behavior may even escalate to physical violence between the males. Scientists have observed this behavior in multiple parts of the world, including the Gulf of St. Lawrence in Canada and the South Taranaki Bight in New Zealand.

Females typically give birth once every two to three years at the start of the winter after a gestation period of 10 to 12 months. The calf weighs about 2.5 tonnes (2.8 short tons) and is around 7 meters (23 ft) in length. Blue whale calves drink 380–570 liters (100–150 U.S. gallons) of milk a day. Blue whale milk has an energy content of about 18,300 kJ/kg (4,370 kcal/kg). The calf is weaned after six months, by which time it has doubled in length. The first video of a calf thought to be nursing was filmed on 5 February 2016.

Sexual maturity is typically reached at five to ten years of age. In the Northern Hemisphere, whaling records show that males averaged 20–21 m (66–69 ft) and females 21–23 m (69–75 ft) at sexual maturity, while in the Southern Hemisphere it was 22.6 and 24 m (74 and 79 ft), respectively. In the Northern Hemisphere, as adults, males averaged 24 m (79 ft) and females 25 m (82 ft) with average calculated weights of 90.5 and 101.5 tonnes (100 and 112 tons), respectively. Blue whales in the eastern North Pacific population were found to be on average 0.91 meters (3.0 feet) shorter, therefore with males averaging 23.3 meters (76 feet) and 80.5 tonnes (88.5 tons) and females 24 meters (79 feet) and 90.5 tonnes (100 tons). Antarctic males averaged 25 m (82 ft) and females 26.2 m (86 ft), averaging 101.5 and 118 tonnes (112 and 130 tons). Pygmy blue whales average 19.2 meters (63 feet) at sexual maturity, with males averaging 21 meters and females 22 meters (69 and 72 feet) when fully grown, averaging 76 and 90 tonnes (83.5 and 99 tons).

In the eastern North Pacific, photogrammetric studies have shown sexually mature (but not necessarily fully grown) blue whales today average 21.7 m (71 ft), and about 65.5 tonnes (72 tons) with the largest found being about 24.5 m (80 ft). A 26.5 m (87 ft) female washed ashore near Pescadero, California in 1979.

The weight of individual blue whales varies significantly according to fat condition. Antarctic blue whales gain 50% of their lean body weight in the summer feeding season, i.e. a blue whale entering the Antarctic weighing 100 tons would leave weighing 150 tons. Pregnant females probably gain 60–65%. The fattened weight is 120% the average weight and the lean weight is 80%.

Scientists estimate that blue whales can live for at least 80 years, but since individual records do not date back into the whaling era, this will not be known with certainty for many years. The longest recorded study of a single individual is 34 years, in the eastern North Pacific.

The whales' only natural predator is the orca. Studies report that as many as 25% of mature blue whales have scars resulting from orca attacks. The mortality rate of such attacks is unknown.

Strandings

Blue whale strandings are extremely uncommon, and, because of the species' social structure, mass strandings are unheard of. When strandings do occur, they can become the focus of public interest. In 1920, a blue whale washed up near Bragar on the Isle of Lewis in the Outer Hebrides of Scotland. It had been shot by whalers, but the harpoon had failed to explode. As with other mammals, the fundamental instinct of the whale was to try to carry on breathing at all costs, even though this meant beaching to prevent itself from drowning. Two of the whale's bones were erected just off a main road on Lewis and remain a tourist attraction.

In June 2015, a female blue whale estimated at 12.2 meters (40 feet) and 20 tonnes (22 tons) was stranded on a beach in Maharashtra, India, the first live stranding in the region. Despite efforts by the Albaug forest department and local fishermen, the whale died 10 hours after being stranded. In August 2009, a wounded blue whale was stranded in a bay in Steingrímsfjördur, Iceland. The first rescue attempt failed, as the whale (thought to be over 20 meters long) towed the >20 ton boat back to shore at speeds of up to 7 miles per hour (11 km/h). The whale was towed to sea after 7 hours by a stronger boat. It is unknown whether it survived. In December 2015, a live blue whale thought to be over 20 meters (66 feet) long was rescued from a beach in Chile. Another stranded blue whale, thought to be about 12.2 meters (40 feet) long, was rescued in India in February 2016. Boats were used in all successful cases.

Vocalizations

Estimates made by Cummings and Thompson suggest the source level of sounds made by blue whales are between 155 and 188 decibels when measured relative to a reference pressure of one micropascal at one meter. All blue whale groups make calls at a fundamental frequency between 10 and 40 Hz; the lowest frequency sound a human can typically perceive is 20 Hz. Blue whale calls last between ten and thirty seconds. Blue whales off the coast of Sri Lanka have been repeatedly recorded making "songs" of four notes, lasting about two minutes each, reminiscent of the well-known humpback whale songs. As this phenomenon has not been seen in any other populations, researchers believe it may be unique to the *B. m. brevicauda* (pygmy) subspecies. The loudest sustained noise from a blue whale was at 188 dB.

The purpose of vocalization is unknown. Richardson *et al.* discuss six possible reasons:

1. Maintenance of inter-individual distance.
2. Species and individual recognition.
3. Contextual information transmission (for example feeding, alarm, courtship).
4. Maintenance of social organization (for example contact calls between females and males).
5. Location of topographic features.
6. Location of prey resources.

Population and Whaling

Hunting Era

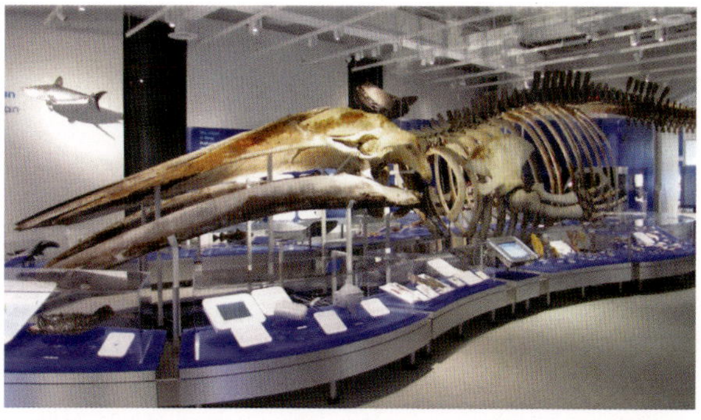

Blue whale skeleton at the Canadian Museum of Nature in Ottawa, Ontario

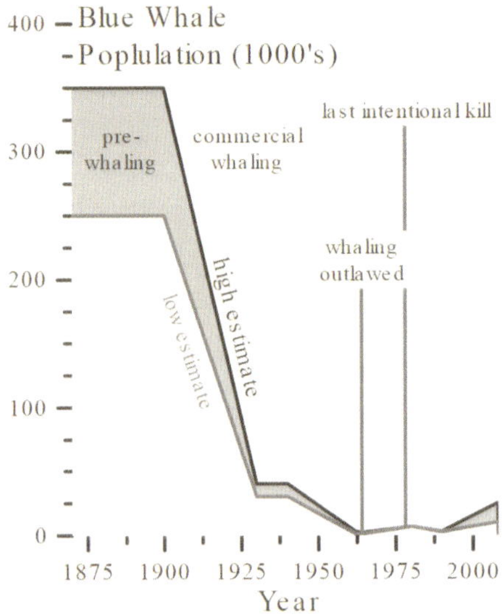

Blue whale populations have declined dramatically due to commercial whaling.

Blue whales are not easy to catch or kill. Their speed and power meant that they were rarely pursued by early whalers, who instead targeted sperm and right whales. In 1864, the Norwegian Svend Foyn equipped a steamboat with harpoons specifically designed for catching large whales. The harpoon gun was initially cumbersome and had a low success rate, but Foyn perfected it, and soon several whaling stations were established on the coast of Finnmark in northern Norway. Because of disputes with the local fishermen, the last whaling station in Finnmark was closed down in 1904.

Soon, blue whales were being hunted off Iceland, the Faroe Islands, Newfoundland, and Spitsbergen. In 1904–05 the first blue whales were taken off South Georgia. By 1925, with the advent of the stern slipway in factory ships and the use of steam-driven whale catchers, the catch of blue whales, and baleen whales as a whole, in the Antarctic and sub-Antarctic began to increase dramatically. In the 1930–31 season, these ships caught 29,400 blue whales in the Antarctic alone. By the end of World War II, populations had been significantly depleted, and, in 1946, the first quotas restricting international trade in whales were introduced, but they were ineffective because of the lack of differentiation between species. Rare species could be hunted on an equal footing with those found in relative abundance.

Arthur C. Clarke, in his 1962 book *Profiles of the Future*, was the first prominent intellectual to call attention to the plight of the blue whale. He mentioned its large brain and said, "we do not know the true nature of the entity we are destroying."

All of the historical coastal Asian groups were driven to near-extinction in short order by Japanese industrial hunts. Those groups that once migrated along western Japan to the East China Sea were likely wiped out much earlier, as the last catches on Amami Oshima were between the 1910s and the 1930s, and the last known stranding records on the Japanese archipelago, excluding the Ryukyu Islands, were over a half-century ago. Commercial catches were continued until 1965 and whaling stations targeting blues were mainly placed along the Hokkaido and Sanriku coasts.

Blue whale hunting was banned in 1966 by the International Whaling Commission, and illegal whaling by the Soviet Union finally halted in the 1970s, by which time 330,000 blue whales had been caught in the Antarctic, 33,000 in the rest of the Southern Hemisphere, 8,200 in the North Pacific, and 7,000 in the North Atlantic. The largest original population, in the Antarctic, had been reduced to a mere 360 individuals, about 0.15% of their initial numbers.

Blue/fin whale hybrids are still caught by the Icelandic whaling industry.

Population and Distribution Today

Since the introduction of the whaling ban, studies have examined whether the conservation reliant global blue whale population is increasing or remaining stable. In the Antarctic, best estimates show an increase of 7.3% per year since the end of illegal Soviet whaling, but numbers remain at under 1% of their original levels. Recovery varies regionally, and the Eastern North Pacific blue whale population (historically a relatively small proportion of the global total) has rebounded to about 2,200 individuals, an estimated 97% of its pre-hunting population.

The total world population was estimated to be between 5,000 and 12,000 in 2002; there are high levels of uncertainty in available estimates for many areas. A more recent estimate by the IUCN puts the global population at 10,000–25,000.

The IUCN Red List counts the blue whale as "endangered", as it has since the list's inception. In the United States, the National Marine Fisheries Service lists them as endangered under the Endangered Species Act. The largest known concentration, consisting of about 2,800 individuals, is the northeast Pacific population of the northern blue whale (*B. m. musculus*) subspecies that ranges from Alaska to Costa Rica, but is most commonly seen from California in summer. Infrequently, this population visits the northwest Pacific between Kamchatka and the northern tip of Japan.

North Atlantic

A blue whale set against the backdrop of the Azores

In the North Atlantic, two stocks of *B. m. musculus* are recognised. The first is found off Greenland, Newfoundland, Nova Scotia and the Gulf of Saint Lawrence. This group is estimated to total about 500. The second, more easterly group is spotted from the Azores in spring to Iceland in July and August; it is presumed the whales follow the Mid-Atlantic Ridge between the two volcanic islands. Beyond Iceland, blue whales have been spotted as far north as Spitsbergen and Jan Mayen, though such sightings are rare. Scientists do not know where these whales spend their winters. The total North Atlantic population is estimated to be between 600 and 1,500. Off Ireland, the first confirmed sightings were made in 2008, since then Porcupine Seabight has been regarded as a prominent habitat for the species along with fin whales. One was sighted along Galicia, Spain, in 2017.

Occasionally blue whales get lost in the Baltic Sea. The remnants of two blue whales have been identified in Finland. One 19-meter skeleton was found on the bottom of the Gulf of Finland during the construction of Nord Stream pipeline. The vertebrae of a second individual were recovered from a field near Pori on the coast of the Bothnian Sea in 1942 and identified in the Copenhagen Zoological Museum by Magnus Degerbøl. Radiocarbon dating revealed the former had wandered into the Littorina Sea over 7000 years ago and perished, while the latter bones were shown to be 800–900 years old.

North Pacific

A blue whale's tail fluke with the Santa Barbara Channel Islands in the background

Five or more subpopulations have been suggested, and several of these mainly in the western North Pacific have been considered either functionally or virtually extinct. Of the populations that once existed off coastal Japan, the last recorded confirmed stranding was in the 1910s. Today, call types suggest only two populations in the North Pacific. Some scientists regard that historical populations off Japan were driven to extinction by whaling activities, mostly from the Kumanonada Sea off Wakayama, in the Gulf of Tosa, and in the Sea of Hyūga. Nowadays, possible vagrants from either eastern or offshore populations are observed on very rare occasions off Kushiro. There were also small, but constant catch records around the Korean Peninsula and in the coastal waters of the Sea of Japan; this species is normally considered not to frequent into marginal seas, such as the Sea of Okhotsk, on usual migrations. Whales were known to migrate further north to eastern Kamchatka, the Gulf of Anadyr, off Abashiri or the southern Sea of Okhotsk, and the Commander Islands. Only three sightings were made between 1994 and 2004 in Russia with one sighted off east coast of the peninsula in 2009, and the last known occurrence in the eastern Sea of Okhotsk was in 1948. In addition, whales have not been confirmed off the Commander Islands for over past 80 years. In 2017, 13 or more whales were observed off Kamchatka and Commander Islands. Historically, wintering grounds existed off the Hawaiian Archipelago, the Northern Mariana Islands, the Bonin Islands and Ryukyu Islands, the Philippines, Taiwan, the Zhoushan Archipelago, and the South China Sea such as in Daya Bay, off the Leizhou Peninsula, and off Hainan Island, and further south to the Paracel Islands. A stranding was recorded in Wanning in 2005. One whale was sighted off Weizhou Island in 2017. For further status in Chinese and Korean waters.

As of 2014, the eastern North Pacific blue whale population had rebounded to an estimated 2,200 individuals, which is thought to be about 97% of its pre-whaling numbers.

Southern Hemisphere and Vicinity to Northern Indian Ocean

Blue whale in Antarctica

In the Southern Hemisphere, there appear to be two distinct subspecies, *B. m. intermedia*, the Antarctic blue whale, and the little-studied pygmy blue whale, *B. m. brevicauda*, found in Indian Ocean waters. The most recent surveys (midpoint 1998) provided an estimate of 2,280 blue whales in the Antarctic (of which fewer than 1% are likely to be pygmy blue whales). Estimates from a 1996 survey show that 424 pygmy blue whales were in a small area south of Madagascar (the Madagascar Plateau) alone, thus it is likely that numbers in the entire Indian Ocean are in the thousands. If this is true, the global numbers would be much higher than estimates predict. However, slower reproduction rate of the species, along with the impacts of whaling, may affect population recoveries as the total population size is predicted to be at less than 50% of its pre-whaling state by 2100.

Several congregating grounds are recently confirmed in Oceania, such as Perth Canyon off Rottnest Island, the Great Australian Bight off Portland, and in South Taranaki Bight and off Kahurangi Point which was discovered just in 2007 and was confirmed in 2014, representing possibly a unique population based on haplotypes. Southern blue and pygmy blue females use waters off Western Australia, and coastal areas of eastern North Island of New Zealand, from Northland waters such as the Bay of Islands and Hauraki Gulf to the Bay of Plenty in the south, as breeding and calving grounds. Whales off southern and western Australia are known to migrate into tropical coastal waters in Indonesia, Philippines, and off East Timor. (Animals in the Philippines may or may not originate from North Pacific populations or from a pygmy blue population in the northern Indian Ocean as whales regularly appear off Bohol, north of the Equator.) At least on occasions, whales also migrate through remote islands such as Cook Islands, and Chilean pelagic waters adjacent to Easter Island and Isla Salas y Gómez, where possibilities of undiscovered wintering grounds have been considered.

Blue whales also migrate through western African waters such as off Angola and Mauritania, and at least whales around Iceland are known to migrate to Mauritania.

Subspecies' Distribution

A fourth subspecies, *B. m. indica*, was identified by Blyth in 1859 in the northern Indian Ocean, but is now thought to be the same subspecies as *B. m. brevicauda*, the pygmy blue whale. Records for Soviet catches seem to indicate that the female adult size is closer to that of the pygmy blue than *B. m. musculus*; the populations of *B. m. indica* and *B. m. brevicauda* appear to be discrete, and the breeding seasons differ by almost six months. Along mainland Indian coasts, appearances of whales had been scarce excluding unconfirmed record, the first blue whale since after the last stranding record in Maharashtra in 1914, was sighted off Kunkeshwar along with several Bryde's whales.

A pygmy blue whale off Sri Lanka

Migratory patterns of these subspecies are not well known. For example, pygmy blue whales have been recorded in the northern Indian Ocean (Oman, Maldives and Sri Lanka), where they may form a distinct resident population. Furthermore, sightings have been recorded from elsewhere in and adjacent to Arabian Sea, including from Gulf of Aden, Persian Gulf, coasts of Bay of Bengal including Bangladesh to Myanmar, and within the Strait of Malacca. The first official confirmation within Thailand's EEZ occurred at Trang.

In addition, the population of blue whales occurring off Chile and Peru may also be a distinct subspecies. Some Antarctic blue whales approach the eastern South Atlantic coast in winter, and occasionally, their vocalizations are heard off Peru, Western Australia, and in the northern Indian Ocean. In Chile, the Cetacean Conservation Center, with support from the Chilean Navy, is undertaking extensive research and conservation work on a recently discovered feeding aggregation of the species off the coast of Chiloe Island in the Gulf of Corcovado (Chiloé National Park), where 326 blue whales were spotted in 2007. In this regions, it is normal for blue whales to enter Fiords. Whales also reach southern Los Lagos, such as off Caleta Zorra, live along with other rorquals.

Efforts to calculate the blue whale population more accurately are supported by marine mammologists at Duke University, who maintain the Ocean Biogeographic Information System—Spatial Ecological Analysis of Megavertebrate Populations (OBIS-SEAMAP), a collation of marine mammal sighting data from around 130 sources.

Threats other than Hunting

Due to their enormous size, power and speed, adult blue whales have virtually no natural predators. There is one documented case in *National Geographic Magazine* of a blue whale being attacked by orcas off the Baja California Peninsula; the orcas were unable to kill the animal outright during their attack, but the blue whale sustained serious wounds and probably died as a result of them shortly after the attack. In March 2014, a pack of orcas harassed a blue whale off California, with one of them biting the tip of the blue whale's tail fluke. The blue whale attempted to tail slap the orca and fled at high speed. A similar incident happened on May 18, 2017 in Monterey Bay, with the orcas swimming in a line up to the blue whale's side. The blue whale fled and escaped. Orcas have virtually no chance against an adult blue whale, but may attack them on occasion anyway for their own enjoyment. Up to a quarter of the blue whales identified in Baja bear scars from orca attacks.

A blue whale surfaces

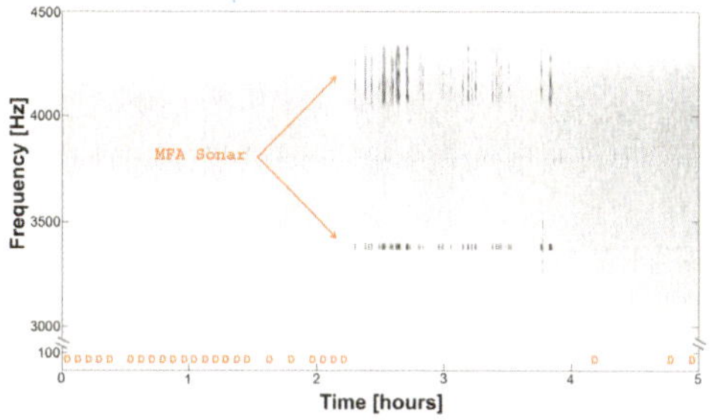

Blue whales stop producing foraging D calls once a mid-frequency sonar is activated, even though the sonar frequency range (1–8 kHz) far exceeds their sound production range (25–100 Hz).

Blue whales may be wounded, sometimes fatally, after colliding with ocean vessels, as well as becoming trapped or entangled in fishing gear. Ship strikes in particular have killed many blue whales in California. In September 2007, three dead blue whales washed up in southern California after being killed by ship strikes. A similar incident occurred in northern California in 2017, possibly also 2010. Ship strikes are also a serious problem in Sri Lanka, and scientists believe this problem could be nearly eliminated by moving the shipping lanes 15 nautical miles to the south. The ever-increasing amount of ocean noise, including sonar, drowns out the vocalizations produced by whales, which makes it harder for them to communicate. Blue whales stop producing foraging D calls once a mid-frequency sonar is activated, even though the sonar frequency range (1–8 kHz) far exceeds their sound production range (25–100 Hz). Research on blue whales in the Southern California Bight has shown that mid-frequency sonar use interferes with foraging behavior, sometimes causing the whales to abandon their feeding. Human threats to the potential recovery of blue whale populations also include accumulation of polychlorinated biphenyl (PCB) chemicals within the whale's body.

With global warming causing glaciers and permafrost to melt rapidly and allowing a large amount of fresh water to flow into the oceans, there are concerns that if the amount of fresh water in the oceans reaches a critical point, there will be a disruption in the thermohaline circulation.

Considering the blue whale's migratory patterns are based on ocean temperature, a disruption in this circulation, which moves warm and cold water around the world, would be likely to have an effect on their migration. The whales summer in the cool, high latitudes, where they feed in krill-abundant waters; they winter in warmer, low latitudes, where they mate and give birth.

The change in ocean temperature would also affect the blue whale's food supply. The warming trend and decreased salinity levels would cause a significant shift in krill location and abundance.

Killer Whale

An orca is a marine mammal. Orcas, or killer whales, are the largest of the dolphins and one of the world's most powerful predators. They feast on marine mammals such as seals, sea lions, and even whales, employing teeth that can be four inches long. They are known to grab seals right off the ice. They also eat fish, squid, and seabirds.

Orcas are incredibly popular as they are the most widely distributed of all whales and dolphins, found in every single ocean! They are very familiar with their black and white coloring but actually, depending on where they live, have very different appearances, behavior, ways of communicating and diet.

Dolphins and whales are closely related. Orcas were given the name 'killer whale' by ancient sailors' observations of groups of orcas hunting and preying on larger whale species. They called orcas asesina ballenas, or 'whale killer' a term that was eventually flipped around to the easier 'killer whale'. Their Latin name, Orcinus orca, also reflects this observation of orcas feeding on large whales. Orcinus translates to 'of the kingdom of the dead,' and orca refers to a kind of whale.

Types

The three to five types of killer whales may be distinct enough to be considered different races, subspecies, or possibly even species. The IUCN reported in 2008, "The taxonomy of this genus is clearly in need of review, and it is likely that *O. orca* will be split into a number of different species or at least subspecies over the next few years." Although large variation in the ecological distinctiveness of different killer whale groups complicate simple differentiation into types, research off the west coast of Canada and the United States in the 1970s and 1980s identified the following three types:

- Resident: These are the most commonly sighted of the three populations in the coastal waters of the northeast Pacific. Residents' diets consist primarily of fish and sometimes squid, and they live in complex and cohesive family groups called pods. Female residents characteristically have rounded dorsal fin tips that terminate in a sharp corner. They visit the same areas consistently. British Columbia and Washington resident populations are amongst the most intensively studied marine mammals anywhere in the world. Researchers have identified and named over 300 killer whales over the past 30 years.
- Transient: The diets of these whales consist almost exclusively of marine mammals. Transients generally travel in small groups, usually of two to six animals, and have less persistent family bonds than residents. Transients vocalize in less variable and less complex dialects. Female transients are characterized by more triangular and pointed dorsal fins than those of residents. The gray or white area around the dorsal fin, known as the "saddle patch", often contains some black colouring in residents. However, the saddle patches of transients are solid and uniformly gray. Transients roam widely along the coast; some individuals have been sighted in both southern Alaska and California. Transients are also referred to as Bigg's killer whale in honor of cetologist Michael Bigg. The term has become increasingly common and may eventually replace the transient label.
- Offshore: A third population of killer whales in the northeast Pacific was discovered in 1988, when a humpback whale researcher observed them in open water. As their name suggests, they travel far from shore and feed primarily on schooling fish. However, because they have large, scarred and nicked dorsal fins resembling those of mammal-hunting transients, it may be that they also eat mammals and sharks. They have mostly been encountered off the west coast of Vancouver Island and near Haida Gwaii. Offshores typically congregate in groups of 20–75, with occasional sightings of larger groups of up to 200. Little is known about their habits, but they are genetically distinct from residents and transients. Offshores appear to be smaller than the others, and females are characterized by dorsal fin tips that are continuously rounded.

Type C killer whales in the Ross Sea in the Southern Ocean: The eye patch slants forward.

Transients and residents live in the same areas, but avoid each other.

Other populations have not been as well studied, although specialized fish and mammal eating killer whales have been distinguished elsewhere. In addition, separate populations of "generalist" (fish- and mammal-eating) and "specialist" (mammal-eating) killer whales have been identified off northwestern Europe. As with residents and transients, the lifestyle of these whales appears to reflect their diet; fish-eating killer whales in Alaska and Norway have resident-like social structures, while mammal-eating killer whales in Argentina and the Crozet Islands behave more like transients.

Three types have been documented in the Antarctic. Two dwarf species, named *Orcinus nanus* and *Orcinus glacialis*, were described during the 1980s by Soviet researchers, but most cetacean researchers are skeptical about their status, and linking these directly to the types described below is difficult.

Some examples of variations in killer whales

- Type A looks like a "typical" killer whale, a large, black-and-white form with a medium-sized white eye patch, living in open water and feeding mostly on minke whales.
- Type B is smaller than type A. It has a large white eye patch. Most of the dark parts of its body are medium gray instead of black, although it has a dark gray patch called a "dorsal cape" stretching back from its forehead to just behind its dorsal fin. The white areas are stained slightly yellow. It feeds mostly on seals.
- Type C is the smallest and lives in larger groups than the others. Its eye patch is distinctively slanted forwards, rather than parallel to the body axis. Like type B, it is primarily white and medium gray, with a dark gray dorsal cape and yellow-tinged patches. Its only observed prey is the Antarctic cod.
- Type D was identified based on photographs of a 1955 mass stranding in New Zealand and six at-sea sightings since 2004. The first video record of this type in life happened between the Kerguelen and Crozet Islands in 2014. It is immediately recognizable by its extremely small white eye patch, narrower and shorter than usual dorsal fin, bulbous head (similar to

a pilot whale), and smaller teeth. Its geographic range appears to be circumglobal in sub-antarctic waters between latitudes 40°S and 60°S. And although nothing is known about the type D diet, it is suspected to include fish because groups have been photographed around longline vessels, where they reportedly prey on Patagonian toothfish (*Dissostichus eleginoides*).

Types B and C live close to the ice pack, and diatoms in these waters may be responsible for the yellowish coloring of both types. Mitochondrial DNA sequences support the theory that these are recently diverged separate species. More recently, complete mitochondrial sequencing indicates the two Antarctic groups that eat seals and fish should be recognized as distinct species, as should the North Pacific transients, leaving the others as subspecies pending additional data. Advanced methods that sequenced the entire mitochondrial genome revealed systematic differences in DNA between different populations.

Mammal-eating killer whales in different regions were long thought likely to be closely related, but genetic testing has refuted this hypothesis.

There are seven identified ecotypes inhabiting isolated ecological niches. Of three orca ecotypes in the Antarctic, one preys on minke whales, the second on seals and penguins, and the third on fish. Another ecotype lives in the eastern North Atlantic, while the three Northeast Pacific ecotypes are labeled the transient, resident and offshore populations described above. Research has supported a proposal to reclassify the Antarctic seal- and fish-eating populations and the North Pacific transients as a distinct species, leaving the remaining ecotypes as subspecies. The first split in the orca population, between the North Pacific transients and the rest, occurred an estimated 700,000 years ago. Such a designation would mean that each new species becomes subject to separate conservation assessments.

Life Cycle

Mother orca with calf near South Georgia

Female killer whales begin to mature at around the age of 10 and reach peak fertility around 20, experiencing periods of polyestrous cycling separated by non-cycling periods of three to 16 months. Females can often breed until age 40, followed by a rapid decrease in fertility. As such, orcas are among the few animals that undergo menopause and live for decades after they have finished breeding. The lifespans of wild females average 50 years. Some are claimed to have lived substantially longer: Granny (J2) was estimated by some researchers to have been as 105 years old at the time of her death, though a biopsy sample indicated her age as 65 to 80 years.

To avoid inbreeding, males mate with females from other pods. Gestation varies from 15 to 18 months. Mothers usually calve a single offspring about once every five years. In resident pods, births occur at any time of year, although winter is the most common. Mortality is extremely high during the first seven months of life, when 37–50% of all calves die. Weaning begins at about 12 months of age, and is complete by two years. According to observations in several regions, all male and female pod members participate in the care of the young.

Males sexually mature at the age of 15, but do not typically reproduce until age 21. Wild males live around 29 years on average, with a maximum of about 60 years. One male, known as Old Tom, was reportedly spotted every winter between the 1840s and 1930 off New South Wales, Australia. This would have made him up to 90 years old. Examination of his teeth indicated he died around age 35, but this method of age determination is now believed to be inaccurate for older animals. One male known to researchers in the Pacific Northwest was estimated to have been 59 years old when he died in 2010. Killer whales are unique among cetaceans, as their caudal sections enlongate with age, making their heads relatively shorter.

Infanticide, once thought to occur only in captive killer whales, was observed in wild populations by researchers off British Columbia on Dec. 2, 2016. In this incident, an adult male killed the calf of a female within the same pod, with his mother also joining in the assault. It is theorized that the male killed the young calf in order to mate with its mother (something that occurs in other carnivore species), while the male's mother supported the breeding opportunity for her son. The attack ended when the calf's mother struck and injured the attacking male. Such behavior matches that of many smaller dolphin species such as the bottlenose dolphin.

Range and Habitat

A killer whale leaps out of the water when swimming—a behavior known as porpoising, in Hood Canal

"Type B" orcas off South Georgia Island

Killer whales are found in all oceans and most seas. Due to their enormous range, numbers, and density, relative distribution is difficult to estimate, but they clearly prefer higher latitudes and coastal areas over pelagic environments. Areas which serve as major study sites for the species include the coasts of Iceland, Norway, the Valdes Peninsula of Argentina, the Crozet Islands, New Zealand and parts of the west coast of North America, from California to Alaska.

Systematic surveys indicate the highest densities of killer whales (>0.40 individuals per 100 km^2) in the northeast Atlantic around the Norwegian coast, in the north Pacific along the Aleutian Islands, the Gulf of Alaska and in the Southern Ocean off much of the coast of Antarctica. They are considered "common" (0.20–0.40 individuals per 100 km^2) in the eastern Pacific along the coasts of British Columbia, Washington and Oregon, in the North Atlantic Ocean around Iceland and the Faroe Islands. High densities have also been reported but not quantified in the western North Pacific around the Sea of Japan, Sea of Okhotsk, Kuril Islands, Kamchatka and the Commander Islands and in the Southern Hemisphere off southern Brazil and the tip of southern Africa. They are reported as seasonally common in the Canadian Arctic, including Baffin Bay between Greenland and Nunavut, as well as Tasmania and Macquarie Island and regularly occurring or distinct populations exist off Northwest Europe, California, Patagonia, the Crozet Islands, Marion Island, southern Australia and New Zealand. The northwest Atlantic population of at least 67 individuals ranges from Labrador and Newfoundland to New England with sightings to Cape Cod and Long Island.

Information for offshore regions and warmer waters is more scarce, but widespread sightings indicate the killer whale can survive in most water temperatures. They have been sighted, though more infrequently, in the Mediterranean, the Arabian Sea, the Gulf of Mexico, and the Caribbean. Over 50 individual whales have been documented in the northern Indian Ocean, including two individuals that were sighted in the Persian Gulf in 2008 and off Sri Lanka in 2015. Those orcas may occasionally enter the Red Sea through the Gulf of Aden. The modern status of the species along coastal mainland China and its vicinity is unknown. Recorded sightings have been made from almost the entire shoreline. A wide-ranging population is likely to exist in the central Pacific, with some sightings off Hawaii. Distinct populations may also exist off the west coast of tropical Africa, and Papua New Guinea. In the Mediterranean, killer whales are considered "visitors", likely from the North Atlantic, and sightings become less frequent further east. However, a small year-round population is known to exist in the Strait of Gibraltar. Killer whales also appear to regularly occur off the Galápagos Islands.

In the Antarctic, killer whales range up to the edge of the pack ice and are believed to venture into the denser pack ice, finding open leads much like beluga whales in the Arctic. However, killer whales are merely seasonal visitors to Arctic waters, and do not approach the pack ice in the summer. With the rapid Arctic sea ice decline in the Hudson Strait, their range now extends deep into the northwest Atlantic. Occasionally, killer whales swim into freshwater rivers. They have been documented 100 mi (160 km) up the Columbia River in the United States. They have also been found in the Fraser River in Canada and the Horikawa River in Japan.

Migration patterns are poorly understood. Each summer, the same individuals appear off the coasts of British Columbia and Washington. Despite decades of research, where these animals go for the rest of the year remains unknown. Transient pods have been sighted from southern Alaska to central California.

Population

Worldwide population estimates are uncertain, but recent consensus suggests a minimum of 50,000. Local estimates include roughly 25,000 in the Antarctic, 8,500 in the tropical Pacific, 2,250–2,700 off the cooler northeast Pacific and 500–1,500 off Norway. Japan's Fisheries Agency estimated 2,321 killer whales were in the seas around Japan.

Feeding

Killer whales are apex predators, meaning that they themselves have no natural predators. They are sometimes called the wolves of the sea, because they hunt in groups like wolf packs. Killer whales hunt varied prey including fish, cephalopods, mammals, sea birds, and sea turtles. Different populations or ecotypes may specialize, and some can have a dramatic impact on prey species. However, whales in tropical areas appear to have more generalized diets due to lower food productivity.

A killer whale skeleton

Fish

Tail-slapping in Vestfjorden, Norway

Resident (fish-eating) killer whales: The curved dorsal fins are typical of resident females.

Fish-eating killer whales prey on around 30 species of fish. Some populations in the Norwegian and Greenland sea specialize in herring and follow that fish's autumnal migration to the Norwegian coast. Salmon account for 96% of northeast Pacific residents' diet, including 65% of large, fatty Chinook. Chum salmon are also eaten, but smaller sockeye and pink salmon are not a significant food item. Depletion of specific prey species in an area is, therefore, cause for concern for local populations, despite the high diversity of prey. On average, a killer whale eats 227 kilograms (500 lb) each day. While salmon are usually hunted by an individual whale or a small group, herring are often caught using carousel feeding: the killer whales force the herring into a tight ball by releasing bursts of bubbles or flashing their white undersides. They then slap the ball with their tail flukes, stunning or killing up to 15 fish at a time, then eating them one by one. Carousel feeding has only been documented in the Norwegian killer whale population, as well as some oceanic dolphin species.

In New Zealand, sharks and rays appear to be important prey, including eagle rays, long-tail and short-tail stingrays, common threshers, smooth hammerheads, blue sharks, basking sharks and shortfin mako sharks. With sharks, orcas may herd them to the surface and strike them with their tail flukes, while bottom-dwelling rays are cornered, pinned to the ground and taken to the surface. In other parts of the world, killer whales have preyed on broadnose sevengill sharks, tiger sharks and even small whale sharks. Killer whales have also been recorded feeding on great white sharks, including one incident filmed near the Farallon Islands in October 1997, where a 4.7–5.3-metre (15–17 ft) female orca killed a 3–4-metre (9.8–13.1 ft) white shark, possibly inducing tonic immobility before feeding. A pod of orcas have been recorded killing a white shark off South Australia, with possible kills in South Africa. The whales appear to target the shark's liver. Competition between killer whales and white sharks is probable in regions where their diets overlap.

Mammals

Killer whales are very sophisticated and effective predators of marine mammals. Thirty-two cetacean species have been recorded as prey, from observing orcas' feeding activity, examining the stomach contents of dead orcas, and seeing scars on the bodies of surviving prey animals. Groups even attack larger cetaceans such as minke whales, gray whales, and, rarely, sperm whales or blue whales.

Hunting a large whale usually takes several hours. Killer whales generally attack young or weak animals; however, a group of five or more may attack a healthy adult. When hunting a young whale, a group chases it and its mother to exhaustion. Eventually, they separate the pair and surround the calf, drowning it by keeping it from surfacing. Pods of female sperm whales sometimes protect themselves by forming a protective circle around their calves with their flukes facing outwards, using them to repel the attackers. Rarely, large killer whale pods can overwhelm even adult female sperm whales. Adult bull sperm whales, which are large, powerful and aggressive when threatened, and fully grown adult blue whales, which are possibly too large to overwhelm, are not believed to be prey for killer whales.

Prior to the advent of industrial whaling, great whales may have been the major food source for killer whales. The introduction of modern whaling techniques may have aided killer whales by the sound of exploding harpoons indicating availability of prey to scavenge, and compressed air inflation of whale carcasses causing them to float, thus exposing them to scavenging. However, the devastation of great whale populations by unfettered whaling has possibly reduced their availability for killer whales, and caused them to expand their consumption of smaller marine mammals, thus contributing to the decline of these as well.

It has been hypothesised that predation by orcas on whale calves in high-productivity, high-latitude areas is the reason for great whale migrations during breeding season to low-productivity tropical waters where orcas are scarcer.

Killer whale hunting a Weddell seal

Other marine mammal prey species include nearly 20 species of seal, sea lion and fur seal. Walruses and sea otters are less frequently taken. Often, to avoid injury, killer whales disable their prey before killing and eating it. This may involve throwing it in the air, slapping it with their tails, ramming it, or breaching and landing on it. Sea lions are killed by head-butting or after a stunning blow from a tail fluke. In the Aleutian Islands, a decline in sea otter populations in the 1990s was controversially attributed by some scientists to killer whale predation, although with no direct evidence. The decline of sea otters followed a decline in harbour seal and Steller sea lion populations, the killer whale's preferred prey, which in turn may be substitutes for their original prey, now decimated by industrial whaling.

In steeply banked beaches off Península Valdés, Argentina, and the Crozet Islands, killer whales feed on South American sea lions and southern elephant seals in shallow water, even beaching temporarily to grab prey before wriggling back to the sea. Beaching, usually fatal to cetaceans, is not an instinctive behaviour, and can require years of practice for the young. Killer whales can then release the animal near juvenile whales, allowing the younger whales to practice the difficult capture technique on the now-weakened prey. "Wave-hunting" killer whales spy-hop to locate Weddell seals, crabeater seals, leopard seals, and penguins resting on ice floes, and then swim in groups to create waves that wash over the floe. This washes the prey into the water, where other killer whales lie in wait.

Killer whales have also been observed preying on terrestrial mammals, such as deer swimming between islands off the northwest coast of North America. Killer whale cannibalism has also been reported based on analysis of stomach contents, but this is likely to be the result of scavenging remains dumped by whalers. One killer whale was also attacked by its companions after being shot. Although resident killer whales have never been observed to eat other marine mammals, they occasionally harass and kill porpoises and seals for no apparent reason.

Birds

Killer whales in many areas may prey on cormorants and gulls. A captive killer whale at Marine Land discovered it could regurgitate fish onto the surface, attracting sea gulls, and then eat the birds. Four others then learned to copy the behavior.

Behavior

Association of killer whales and black-browed albatrosses

Day-to-day killer whale behavior generally consists of foraging, travelling, resting and socializing. Killer whales frequently engage in surface behavior such as breaching (jumping completely out of the water) and tail-slapping. These activities may have a variety of purposes, such as courtship, communication, dislodging parasites, or play. Spyhopping is a behavior in which a whale holds its head above water to view its surroundings.

Resident killer whales swim alongside porpoises, other dolphins, seals, and sea lions, which are common prey for transient killer whales.

Social Structure

A pair of killer whales in a transient population in Northwest Pacific

Killer whales are notable for their complex societies. Only elephants and higher primates live in comparably complex social structures. Due to orcas' complex social bonds, many marine experts have concerns about how humane it is to keep them in captivity.

Resident killer whales in the eastern North Pacific live in particularly complex and stable social groups. Unlike any other known mammal social structure, resident whales live with their mothers for their entire lives. These family groups are based on matrilines consisting of the eldest female (matriarch) and her sons and daughters, and the descendants of her daughters, etc. The average size of a matriline is 5.5 animals. Because females can reach age 90, as many as four generations travel together. These matrilineal groups are highly stable. Individuals separate for only a few hours at a time, to mate or forage. With one exception, a killer whale named Luna, no permanent separation of an individual from a resident matriline has been recorded.

Killer whales, like this one near Alaska, commonly breach, often lifting their entire bodies out of the water.

Closely related matrilines form loose aggregations called pods, usually consisting of one to four matrilines. Unlike matrilines, pods may separate for weeks or months at a time. DNA testing indicates resident males nearly always mate with females from other pods. Clans, the next level of resident social structure, are composed of pods with similar dialects, and common but older maternal heritage. Clan ranges overlap, mingling pods from different clans. The final association layer, perhaps more arbitrarily defined than the familial groupings, is called the community, and is defined as a set of clans that regularly commingle. Clans within a community do not share vocal patterns.

Transient pods are smaller than resident pods, typically consisting of an adult female and one or two of her offspring. Males typically maintain stronger relationships with their mothers than other females. These bonds can extend well into adulthood. Unlike residents, extended or permanent separation of transient offspring from natal matrilines is common, with juveniles and adults of both sexes participating. Some males become "rovers" and do not form long-term associations, occasionally joining groups that contain reproductive females. As in resident clans, transient community members share an acoustic repertoire, although regional differences in vocalizations have been noted.

Vocalizations

Like all cetaceans, killer whales depend heavily on underwater sound for orientation, feeding, and communication. They produce three categories of sounds: clicks, whistles, and pulsed calls. Clicks are believed to be used primarily for navigation and discriminating prey and other objects in the surrounding environment, but are also commonly heard during social interactions.

Northeast Pacific resident groups tend to be much more vocal than transient groups in the same waters. Residents feed primarily on Chinook and chum salmon, species that are insensitive to killer whale calls (inferred from the audio-gram of Atlantic salmon). In contrast, the marine mammal prey of transients hear well underwater at the frequencies used in killer whale calls. As such, transients are typically silent, probably to avoid alerting their mammalian prey. They sometimes use a single click (called a cryptic click) rather than the long train of clicks observed in other populations. Residents are only silent when resting.

All members of a resident pod use similar calls, known collectively as a dialect. Dialects are composed of specific numbers and types of discrete, repetitive calls. They are complex and stable over time. Call patterns and structure are distinctive within matrilines. Newborns produce calls similar to their mothers, but have a more limited repertoire. Individuals likely learn their dialect through contact with their mother and other pod members. For instance, family-specific calls have been observed more frequently in the days following a calf's birth, which may help the calf learn them. Dialects are probably an important means of maintaining group identity and cohesiveness. Similarity in dialects likely reflects the degree of relatedness between pods, with variation building over time. When pods meet, dominant call types decrease and subset call types increase. The use of both call types is called biphonation. The increased subset call types may be the distinguishing factor between pods and inter-pod relations.

Dialects of killer whales not only distinguish them between pods, but also between types. Resident dialects contain seven to 17 (mean = 11) distinctive call types. All members of the North American west coast transient community express the same basic dialect, although minor regional variation in call types is evident. Preliminary research indicates offshore killer whales have group-specific dialects unlike those of residents and transients.

The vocalizations of killer whales in other parts of the world have also been studied. Norwegian and Icelandic herring-eating orcas appear to have different vocalizations for activities like hunting and traveling.

Intelligence

Killer whales have the second-heaviest brains among marine mammals (after sperm whales, which have the largest brain of any animal). They can be trained in captivity and are often described as intelligent, although defining and measuring "intelligence" is difficult in a species whose environment and behavioral strategies are very different from those of humans.

A killer whale plays with a ball of ice, soon after a researcher had thrown a snowball at the whale.

Killer whales imitate others, and seem to deliberately teach skills to their kin. Off the Crozet Islands, mothers push their calves onto the beach, waiting to pull the youngster back if needed.

People who have interacted closely with killer whales offer numerous anecdotes demonstrating the whales' curiosity, playfulness, and ability to solve problems. Alaskan killer whales have not only learned how to steal fish from longlines, but have also overcome a variety of techniques designed to stop them, such as the use of unbaited lines as decoys. Once, fishermen placed their boats several miles apart, taking turns retrieving small amounts of their catch, in the hope that the whales would not have enough time to move between boats to steal the catch as it was being retrieved. A researcher described what happened next:

> It worked really well for a while. Then the whales split into two groups. It didn't even take them an hour to figure it out. They were so thrilled when they figured out what was going on, that we were playing games. They were breaching by the boats.
>
> —*Craig Matkin*

In other anecdotes, researchers describe incidents in which wild killer whales playfully tease humans by repeatedly moving objects the humans are trying to reach, or suddenly start to toss around a chunk of ice after a human throws a snowball.

The killer whale's use of dialects and the passing of other learned behaviours from generation to generation have been described as a form of animal culture.

The complex and stable vocal and behavioural cultures of sympatric groups of killer whales (*Orcinus orca*) appear to have no parallel outside humans and represent an independent evolution of cultural faculties.

(Two species or populations are considered sympatric when they live in the same geographic area and thus regularly encounter one another.)

Conservation

The type C killer whale has two-toned gray colouring, including a dark "dorsal cape", in body areas where most killer whales have solid black colouring.

In 2008, the IUCN (International Union for Conservation of Nature) changed its assessment of the killer whale's conservation status from conservation dependent to data deficient, recognizing that one or more killer whale types may actually be separate, endangered species. Depletion of prey species, pollution, large-scale oil spills, and habitat disturbance caused by noise and conflicts with boats are the most significant worldwide threats.

Like other animals at the highest trophic levels, the killer whale is particularly at risk of poisoning from bioaccumulation of toxins, including polychlorinated biphenyls (PCBs). European harbor seals have problems in reproductive and immune functions associated with high levels of PCBs and related contaminants, and a survey off the Washington coast found PCB levels in killer whales were higher than levels that had caused health problems in harbor seals. Blubber samples in the Norwegian Arctic show higher levels of PCBs, pesticides and brominated flame-retardants than in polar bears. When food is scarce, killer whales metabolize blubber for energy, which increases pollutant concentrations in their blood.

In the Pacific Northwest, wild salmon stocks, a main resident food source, have declined dramatically in recent years. In the Puget Sound region only 75 whales remain with few births over the last few years. On the west coast of Alaska and the Aleutian Islands, seal and sea lion populations have also substantially declined.

In 2005, the United States government listed the southern resident community as an endangered population under the Endangered Species Act. This community comprises three pods which live mostly in the Georgia and Haro Straits and Puget Sound in British Columbia and Washington. They do not breed outside of their community, which was once estimated at around 200 animals and later shrank to around 90. In October 2008, the annual survey revealed seven were missing and presumed dead, reducing the count to 83. This is potentially the largest decline in the population in the past ten years. These deaths can be attributed to declines in Chinook salmon.

An adult female and her calf

Scientist Ken Balcomb has extensively studied killer whales since 1976; he is the research biologist responsible for discovering U.S. Navy sonar may harm killer whales. He studied killer whales from the Center for Whale Research, located in Friday Harbor, Washington. He was also able to study killer whales from "his home porch perched above Puget Sound, where the animals hunt and play in summer months". In May 2003, Balcomb (along with other whale watchers near the Puget Sound coastline) noticed uncharacteristic behaviour displayed by the killer whales. The whales seemed "agitated and were moving haphazardly, attempting to lift their heads free of the water" to escape the sound of the sonars. "Balcomb confirmed at the time that strange underwater pinging noises detected with underwater microphones were sonar. The sound originated from a U.S. Navy frigate 12 miles (19 kilometers) distant, Balcomb said." The impact of sonar waves on killer whales is potentially life-threatening. Three years prior to Balcomb's discovery, research in the Bahamas showed 14 beaked whales washed up on the shore. These whales were beached on the day U.S. Navy destroyers were activated into sonar exercise. Of the 14 whales beached, six of them died. These six dead whales were studied, and CAT scans of two of the whale heads showed hemorrhaging around the brain and the ears, which is consistent with decompression sickness.

Another conservation concern was made public in September 2008 when the Canadian government decided it was not necessary to enforce further protections (including the Species at Risk Act in place to protect endangered animals along their habitats) for killer whales aside from the laws already in place. In response to this decision, six environmental groups sued the federal government, claiming killer whales were facing many threats on the British Columbia Coast and the federal government did nothing to protect them from these threats. A legal and scientific nonprofit organization, Ecojustice, led the lawsuit and represented the David Suzuki Foundation, Environmental Defence, Greenpeace Canada, International Fund for Animal Welfare, the Raincoast Conservation Foundation, and the Wilderness Committee. Many scientists involved in this lawsuit, including Bill Wareham, a marine scientist with the David Suzuki Foundation, noted increased boat traffic, water toxic wastes, and low salmon population as major threats, putting approximately 87 killer whales on the British Columbia Coast in danger.

The last known AT1 pod offspring, AT3, swimming in Resurrection Bay

Underwater noise from shipping, drilling, and other human activities is a significant concern in some key killer whale habitats, including Johnstone Strait and Haro Strait. In the mid-1990s, loud underwater noises from salmon farms were used to deter seals. Killer whales also avoided the surrounding waters. High-intensity sonar used by the Navy disturbs killer whales along with other marine mammals. Killer whales are popular with whale watchers, which may stress the whales and alter their behavior, particularly if boats approach too closely or block their lines of travel.

The *Exxon Valdez* oil spill adversely affected killer whales in Prince William Sound and Alaska's Kenai Fjords region. Eleven members (about half) of one resident pod disappeared in the following year. The spill damaged salmon and other prey populations, which in turn damaged local killer whales. By 2009, scientists estimated the AT1 transient population (considered part of a larger population of 346 transients), numbered only seven individuals and had not reproduced since the spill. This population is expected to die out.

Relationship with Humans

Indigenous Cultures

Haida sculpture by Bill Reid

The indigenous peoples of the Pacific Northwest Coast feature killer whales throughout their art, history, spirituality and religion. The Haida regarded killer whales as the most powerful animals in the ocean, and their mythology tells of killer whales living in houses and towns under the sea. According to these myths, they took on human form when submerged, and humans who drowned went to live with them. For the Kwakwaka'wakw, the killer whale was regarded as the ruler of the undersea world, with sea lions for slaves and dolphins for warriors. In Nuu-chah-nulth and Kwakwaka'wakw mythology, killer whales may embody the souls of deceased chiefs. The Tlingit of southeastern Alaska regarded the killer whale as custodian of the sea and a benefactor of humans.

The Maritime Archaic people of Newfoundland also had great respect for killer whales, as evidenced by stone carvings found in a 4,000-year-old burial at the Port au Choix Archaeological Site.

In the tales and beliefs of the Siberian Yupik people, killer whales are said to appear as wolves in winter, and wolves as killer whales in summer. Killer whales are believed to assist their hunters in driving walrus. Reverence is expressed in several forms: the boat represents the animal, and a wooden carving hung from the hunter's belt. Small sacrifices such as tobacco are strewn into the sea for them. Killer whales were believed to have helped the hunters even when in wolf guise, by forcing reindeer to allow themselves to be killed.

Indigenous Ainu tribe often refereed killer whales in their folklore and myth as *Repun Kamuy* (God of Sea/Offshore) to bring fortunes (whales) to the coasts, and there had been traditional funerals for stranded or deceased orcas akin to funerals for other animals such as brown bears.

"Killer" Stereotype

In Western cultures, killer whales were historically feared as dangerous, savage predators. The first written description of a killer whale was given by Pliny the Elder *circa* AD 70, who wrote, "Orcas (the appearance of which no image can express, other than an enormous mass of savage flesh with teeth) are the enemy of other whales they charge and pierce them like warships ramming."

Male killer whale depicted in St Mary's in Greifswald

Of the very few confirmed attacks on humans by wild killer whales, none have been fatal. In one instance, killer whales tried to tip ice floes on which a dog team and photographer of the Terra Nova Expedition were standing. The sled dogs' barking is speculated to have sounded enough like seal calls to trigger the killer whale's hunting curiosity. In the 1970s, a surfer in California was bitten, and in 2005, a boy in Alaska who was splashing in a region frequented by harbor seals was bumped by a killer whale that apparently misidentified him as prey. Unlike wild killer whales, captive killer whales are reported to have made nearly two dozen attacks on humans since the 1970s, some of which have been fatal.

Competition with fishermen also led to killer whales being regarded as pests. In the waters of the Pacific Northwest and Iceland, the shooting of killer whales was accepted and even encouraged by governments. As an indication of the intensity of shooting that occurred until fairly recently, about 25% of the killer whales captured in Puget Sound for aquarium through 1970 bore bullet scars. The U.S. Navy claimed to have deliberately killed hundreds of killer whales in Icelandic waters in 1956 with machine-guns, rockets, and depth charges.

Modern Western Attitudes

Ingrid Visser's research team filming orcas in New Zealand

Western attitudes towards killer whales have changed dramatically in recent decades. In the mid-1960s and early 1970s, killer whales came to much greater public and scientific awareness, starting with the first live-capture and display of a killer whale known as Moby Doll, a resident harpooned off Saturna Island in 1964. So little was known at the time, it was nearly two months before the whale's keepers discovered what food (fish) it was willing to eat. To the surprise of those who saw him, Moby Doll was a docile, nonaggressive whale that made no attempts to attack humans.

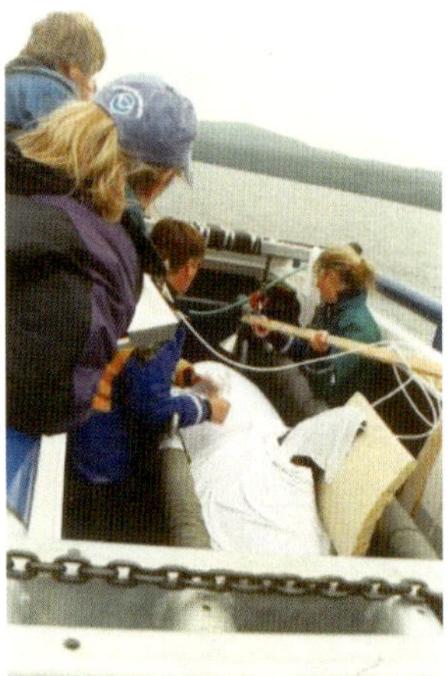

The orphan Springer was successfully returned to her family.

Between 1964 and 1976, 50 killer whales from the Pacific Northwest were captured for display in aquaria, and public interest in the animals grew. In the 1970s, research pioneered by Michael Bigg led to the discovery of the species' complex social structure, its use of vocal communication, and its extraordinarily stable mother–offspring bonds. Through photo-identification techniques, individuals were named and tracked over decades.

Bigg's techniques also revealed the Pacific Northwest population was in the low hundreds rather than the thousands that had been previously assumed. The southern resident community alone had lost 48 of its members to captivity; by 1976, only 80 remained. In the Pacific Northwest, the species that had unthinkingly been targeted became a cultural icon within a few decades.

The public's growing appreciation also led to growing opposition to whale–keeping in aquarium. Only one whale has been taken in North American waters since 1976. In recent years, the extent of the public's interest in killer whales has manifested itself in several high-profile efforts surrounding individuals. Following the success of the 1993 film *Free Willy,* the movie's captive star Keiko was returned to the coast of his native Iceland in 1998. The director of the International Marine Mammal Project for the Earth Island Institute, David Phillips, led the efforts to return Keiko to the Iceland waters. In 2002, the orphan Springer was discovered in Puget Sound, Washington. She became the first whale to be successfully reintegrated into a wild pod after human intervention, crystallizing decades of research into the vocal behavior and social structure of the region's killer whales. The saving of Springer raised hopes that another young killer whale named Luna, which had become separated from his pod, could be returned to it. However, his case was marked by controversy about whether and how to intervene, and in 2006, Luna was killed by a boat propeller.

Whaling

The killer whale named Old Tom swims alongside a whaleboat, flanking a whale calf. The boat is being towed by a harpooned whale (not visible here), near Eden, Australia.

The earlier of known records of commercial hunting of killer whales date to the 18th century in Japan. During the 19th and early 20th centuries, the global whaling industry caught immense numbers of baleen and sperm whales, but largely ignored killer whales because of their limited amounts of recoverable oil, their smaller populations, and the difficulty of taking them. Once the stocks of larger species were depleted, killer whales were targeted by commercial whalers in the mid-20th century. Between 1954 and 1997, Japan took 1,178 killer whales (although the Ministry of the Environment claims that there had been domestic catches of about 1,600 whales between late 1940s to 1960s) and Norway took 987. Over 3,000 killer whales were taken by Soviet whalers, including an Antarctic catch of 916 in 1979–80 alone, prompting the International Whaling Commission to recommend a ban on commercial hunting of the species pending further research. (Compare with the situation on land, commercial hunting.) Today, no country carries out a substantial hunt, although Indonesia and Greenland permit small subsistence hunts. Other than commercial hunts, killer whales were hunted along Japanese coasts out of public concern for potential conflicts with fisheries. Such cases include a semi-resident male-female pair in Akashi Strait and

Harimanada being killed in the Seto Inland Sea in 1957, the killing of five whales from a pod of 11 members that swam into Tokyo Bay in 1970, and a catch record in southern Taiwan in the 1990s.

Cooperation with Humans

Killer whales have helped humans hunting other whales. One well-known example was the killer whales of Eden, Australia, including the male known as Old Tom. Whalers more often considered them a nuisance, however, as orcas would gather to scavenge meat from the whalers' catch. Some populations, such as in Alaska's Prince William Sound, may have been reduced significantly by whalers shooting them in retaliation.

Captivity

Lolita, at the Miami Seaquarium, is one of the oldest whales in captivity.

The killer whale's intelligence, trainability, striking appearance, playfulness in captivity and sheer size have made it a popular exhibit at aquaria and aquatic theme parks. From 1976 to 1997, 55 whales were taken from the wild in Iceland, 19 from Japan, and three from Argentina. These figures exclude animals that died during capture. Live captures fell dramatically in the 1990s, and by 1999, about 40% of the 48 animals on display in the world were captive-born.

Organizations such as World Animal Protection and the Whale and Dolphin Conservation Society campaign against the practice of keeping them in captivity. In captivity, they often develop pathologies, such as the dorsal fin collapse seen in 60–90% of captive males. Captives have vastly reduced life expectancies, on average only living into their 20s. In the wild, females who survive infancy live 46 years on average, and up to 70–80 years in rare cases. Wild males who survive infancy live 31 years on average, and up to 50–60 years. Captivity usually bears little resemblance to wild habitat, and captive whales' social groups are foreign to those found in the wild. Critics claim captive life is stressful due to these factors and the requirement to perform circus tricks that are not part of wild killer whale behavior, see above. Wild killer whales may travel up to 160 kilometres (100 mi) in a day, and critics say the animals are too big and intelligent to be suitable for captivity. Captives occasionally act aggressively towards themselves, their tankmates, or humans, which critics say is a result of stress. Between 1991 and 2010, the bull orca known as Tilikum was involved in the death of three people, and was featured in the critically acclaimed 2013 film, *Blackfish*. Tilikum lived at SeaWorld from 1992 until his death in 2017.

A 2015 study coauthored by staff at SeaWorld and the Minnesota Zoo indicates that there is no significant difference in survivorship between free-ranging and captive killer whales. The authors speculate about the future utility of studying captive populations for the purposes of understanding orca biology and the implications of such research of captive animals in the overall health of both wild and marine park populations.

As of March 2016, SeaWorld has announced that they will be ending their orca breeding program and their theatrical shows. They previously announced, in November 2015, that the shows would be coming to an end in San Diego but it is now to happen in both Orlando and San Antonio as well.

Amazon River Dolphin

The Amazon River dolphin, also known as the pink river dolphin or boto, lives only in freshwater. It is found throughout much of the Amazon and Orinoco river basins in Bolivia, Brazil, Colombia, Ecuador, Guyana, Peru, and Venezuela. It is a relatively abundant freshwater cetacean with an estimated population in the tens of thousands. However, it is classified as vulnerable in certain areas due to dams that fragment and threaten certain populations, and from other threats such as contamination of rivers and lakes.

This water mammal is the largest of all the other river dolphin species, and cannot survive in salt water. It is a very distant relative to the sea dolphin, and has a pink to pinkish-grey pigmentation on their bodies, which is one of their primary characteristics, giving them a very unique appearance (and hence their name).

Physical Characteristics

- Size: Length of the adult dolphins varies between 2 and 3 meters.
- Weight: Can weigh up to 160 kilos.
- Skin Color: Color is dependent on their age. While the baby Amazon river dolphins have a dark gray dorsal side, the ventral side is lighter gray. With maturity, the flanks and the ventral side begins turning pink. Pink is predominant in the males. Their colors fade gradually with age, having blue-gray and white tinges all over.

- Eyes: Small round eyes but with a strong eyesight.
- Beaks: The beaks are pretty long with a strong built adapted to capture prey.
- Teeth: The jawline is made of a combination of 24-34 conical and molar-like teeth.
- Fins: All fins are same like other dolphins except that they don't have dorsal fins, but instead, has a hump on the back that gets larger with age.

Longevity

Apure the dolphin lived for forty years at the Duisburg Zoo

Life expectancy of the Amazon river dolphin in the wild is unknown, but in captivity, the longevity of healthy individuals has been recorded at between ten and thirty years. However, the average longevity in captive animals is only 33 months. An individual named Apure at Duisburg Zoo, Germany, is estimated to have lived more than forty years, being captured from the wild at age nine.

Behavior

The Amazon river dolphin are commonly seen singly or in twos, but may also occur in pods that rarely contain more than eight individuals. Pods as large as 37 individuals have been seen in the Amazon, but average is three. In the Orinoco, the largest observed groups number 30, but average is just above five. Typically, social bonds occur between mother and child, but may also been seen in heterogeneous groups or bachelor groups. The largest congregations are seen in areas with abundant food, and at the mouths of rivers. There is significant segregation during the rainy season, with males occupying the river channels, while females and their offspring are located in flooded areas. However, in the dry season, there is no such separation. Due to the high level of prey fish, larger group-sizes are seen in large sections that are directly influenced by white water (such as main rivers and lakes, especially during low water season) than in smaller sections influenced by black water (such as channels and smaller tributaries). In their freshwater habitat they are apex predators and gatherings depend more on food sources and habitat availability than in oceanic dolphins where protection from larger predators is necessary.

Captive studies have shown that the Amazon river dolphin is less shy than the bottlenose dolphin, but also less sociable. It is very curious and has a remarkable lack of fear of foreign objects.

However, dolphins in captivity may not show the same behavior that they do in their natural environment, where they have been reported to hold the oars of the fishermen, rub against the boat, pluck underwater plants, and play with sticks, logs, clay, turtles, snakes, and fish.

They are slow swimmers; they commonly travel at speeds of 1.5 to 3.2 kilometers per hour (0.93 to 1.99 mph) but have been recorded to swim at speeds up to 14 to 22 kilometers per hour (8.7 to 13.7 mph). When they surface, the tips of the snout, melon and dorsal fins appear simultaneously, the tail rarely showing before diving. They can also shake their fins, and pull their tail fin and head above the water to observe the environment. They occasionally jump out of the water, sometimes as high as a meter (3.14 ft). They are harder to train than most other species of dolphin.

Courtship

Adult males have been observed carrying objects in their mouths such as branches or other floating vegetation, or balls of hardened clay. The males appear to carry these objects as a socio-sexual display which is part of their mating system. The behavior is "triggered by an unusually large number of adult males and/or adult females in a group, or perhaps it attracts such into the group. A plausible explanation of the results is that object-carrying is aimed at females and is stimulated by the number of females in the group, while aggression is aimed at other adult males and is stimulated by object-carrying in the group." Before determining that the species had an evident sexual dimorphism, it was postulated that the river dolphins were monogamous. Later, it was shown that males were larger than females and are documented wielding an aggressive sexual behavior in the wild and in captivity. Males often have a significant degree of damage in the dorsal, caudal, and pectoral fins, as well as the blowhole, due to bites and abrasions. They also commonly have numerous secondary teeth-raking scars. This suggests fierce competition for access to females, with a polygynous mating system, though polyandry and promiscuity cannot be excluded.

In captivity, courtship and mating foreplay have been documented. The male takes the initiative by nibbling the fins of the female, but reacts aggressively if the female is not receptive. A high frequency of copulations in a couple was observed; they used three different positions: contacting the womb at right angles, lying head to head, or head to tail.

Reproduction

Breeding is seasonal, and births occur between May and June. The period of birthing coincides with the flood season, and this may provide an advantage because the females and their offspring remain in flooded areas longer than males. As the water level begins to decrease, the density of food sources in flooded areas increases due to loss of space, providing enough energy for infants to meet the high demands required for growth. Gestation is estimated to be around eleven months and captive births take 4 to 5 hours. At birth, calves are 80 centimeters (31 in) long and in captivity have registered a growth of 0.21 meters (0.69 ft) per year. Lactation takes about a year. The interval between births is estimated between 15 and 36 months, and the young dolphins are thought to become independent within two to three years.

The relatively long duration of breastfeeding and parenting suggests a strong mother-child bond. Most couples observed in their natural environment consist of a female and her calf. This suggests that long periods of parental care contribute to learning and development of the young.

Diet

Amazon river dolphin feeding

The diet of the Amazon river dolphin is the most diverse of the toothed whales. It consists of at least 53 different species of fish, grouped in 19 families. The prey size is between 5 and 80 centimetres (2.0 and 31.5 in), with an average of 20 centimetres (7.9 in). The most frequently consumed fish belong to the families Sciaenidae (croakers), Cichlidae, and Characidae (tetras and piranhas). The dolphin's dentition allows it to access shells of river turtles and freshwater crabs. The diet is more diverse during the wet season, when fish are spread in flooded areas outside riverbeds, thus becoming more difficult to catch. The diet becomes more selective during the dry season when prey density is greater.

Usually, these dolphins are active and feeding throughout the day and night. However, they are predominantly crepuscular. They consume about 5.5% of their body weight per day. They sometimes take advantage of the disturbances made by boats to catch disoriented prey. Sometimes, they associate with the distantly-related tucuxi (*Sotalia fluviatilis*), and giant otters (*Pteronura brasiliensis*) to hunt in a coordinated manner, by gathering and attacking fish stocks at the same time. Apparently, there is little competition for food between these species, as each prefers different prey. It has also been observed that captive dolphins share food.

Echolocation

Amazonian rivers are often very murky, and the Amazon river dolphin is therefore likely to depend much more on its sense of echolocation than vision when navigating and finding prey. However, echolocation in shallow waters and flooded forests may result in many echoes to keep track of. For each click produced a multitude of echoes are likely to return to the echolocating animal almost on top of each other which makes object discrimination difficult. This may be why the Amazon river dolphin produces less powerful clicks compared to other similar sized toothed whales. By sending out clicks of lower amplitude only nearby objects will cast back detectable echoes, and hence fewer echoes need to be sorted out, but the cost is a reduced biosonar range. Toothed whales generally do not produce a new echolocation click until all relevant echoes from the previous click have been received, so if detectable echoes are only reflected back from nearby objects, the echoes will quickly return, and the Amazon river dolphin is then able to click at a high rate. This in turn allow these animals to have a high acoustic update rate on their surroundings which may aid prey tracking when echolocating in shallow rivers and flooded forests with plenty of hiding places for the prey.

Communication

Like other dolphins, river dolphins use whistling tones to communicate. The issuance of these sounds is related to the time they return to the surface before diving, suggesting a link to food. Acoustic analysis revealed that the vocalizations are different in structure from the typical whistles of other species of dolphins.

Distribution and Population

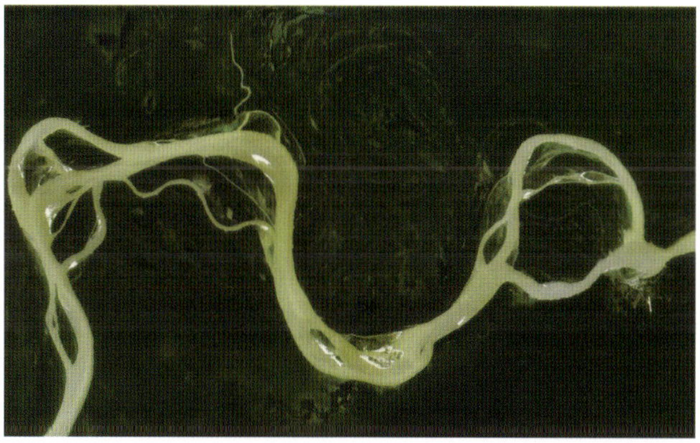

The main branch of the Amazon River near Fonte Boa, Brazil, with multiple floodplains, lakes and smaller channels. The Amazon river dolphin is observed here throughout the year

Amazon river dolphins are the most widespread river dolphins. They are present in six countries in South America: Bolivia, Brazil, Colombia, Ecuador, Peru, and Venezuela, in an area covering about 7,000,000 square kilometres (2,700,000 sq mi). The boundaries are set by waterfalls, such as the Xingu and Tapajós rivers in Brazil, as well as very shallow water. A series of rapids and waterfalls in the Madeira River have isolated one population, recognized as the subspecies *I. g. boliviensis*, in the southern part of the Amazon basin in Bolivia.

They are also distributed in the basin of the Orinoco River, except the Caroni River and the upper Caura River in Venezuela. The only connection between the Orinoco and the Amazon is through the Casiquiare canal. The distribution of dolphins in the rivers and surrounding areas depends on the time of year; in the dry season they are located in the river beds, but in the rainy season, when the rivers overflow, they disperse to the flooded areas, both the forests and the plains.

Studies to estimate the population are difficult to analyze due to the difference in the methodology used. In a study conducted in the stretch of the Amazon called Solimões River, with a length of 1,200 kilometers (750 mi) between the cities of Manaus and Tabatinga, a total of 332 individuals was sighted ± 55 per inspection. Density was estimated at 0.08- 0.33 animals per square kilometer in the main channels, and 0.49 to 0.93 animals per square kilometer in the branches. In another study, on a stretch of 120 kilometers (75 mi) at the confluence of Colombia, Brazil and Peru, 345 individuals with a density of 4.8 per square km in the tributaries around the islands. 2.7 and 2.0 were observed along the banks. Additionally, another study was conducted in the Amazon at the height of the mouth of the Caqueta River for six days. As a result of the studies conducted, it was found that the density is higher in the riverbanks, 3.7 per km, decreasing towards the center of the river. In studies conducted during the rainy season, the density observed in the flood plain was 18 animals per square km, while on the banks of rivers and lakes ranged from 1.8 to 5.8 individuals per square km. These observations suggest that the Amazon river dolphin is found in higher density than any other cetacean.

Habitat

The Amazon river dolphin is located in most of the area's aquatic habitats, including; river basins, major courses of rivers, canals, river tributaries, lakes, and at the ends of rapids and waterfalls. Cyclical changes in the water levels of rivers take place throughout the year. During the dry season, dolphins occupy the main river channels, and during the rainy season, they can move easily to smaller tributaries, to the forest, and to floodplains.

Males and females appear to have selective habitat preferences, with the males returning to the main river channels when water levels are still high, while the females and their offspring remain in the flooded areas as long as possible; probably because it decreases the risk of aggression by males toward the young and predation by other species.

Migration

In the Pacaya-Samiria National Reserve, Peru, photo-identification is used to recognize individuals based on pigmentation patterns, scars and abnormalities in the beak. 72 individuals were recognized, of which 25 were again observed between 1991 and 2000. The intervals between sightings ranged from one day to 7.5 years. The maximum range of motion was 220 kilometers (140 mi), with an average of 60.8 kilometers (37.8 mi). The longest distance in one day was 120 kilometers (75 mi), with an average of 14.5 kilometers (9.0 mi). In a previous study conducted at the center of the Amazon River, a dolphin was observed that moved only a few dozen kilometers from the dry season and wet season. However, three of the reviewed 160 animals were observed over 100 kilometers (62 mi) from where they were first registered. Research in 2011 concluded that photo-identification by skilled operatives using high-quality digital equipment could be a useful tool in monitoring population size, movements and social patterns.

Interactions with Humans

In Captivity

A trained Amazon river dolphin at the Acuario de Valencia

The Amazon river dolphin has historically been kept in dolphinariums. Today, only three exist in captivity: one in Acuario de Valencia in Venezuela, one in Zoologico de Guistochoca in Peru, and one in Duisburg Zoo in Germany. Several hundred were captured between the 1950s and 1970s, and were distributed in dolphinariums throughout the US, Europe, and Japan. Around 100 went to US dolphinariums, and of that, only 20 survived; the last died at the Pittsburgh Zoo in 2002.

Threats

The region of the Amazon in Brazil has an extension of 5,000,000 km² (1,900,000 sq mi) containing diverse fundamental ecosystems. One of these ecosystems is a floodplain, or a várzea forest, and is home to a large number of fish species which are an essential resource for human consumption. The várzea is also a major source of income through excessive local commercialized fishing. Várzea consists of muddy river waters containing a vast number and diversity of nutrient rich species. The abundance of distinct fish species lures the Amazon River dolphin into the várzea areas of high water occurrences during the seasonal flooding.

In addition to attracting predators such as the Amazon river dolphin, these high-water occurrences are an ideal location to draw in the local fisheries. Human fishing activities directly compete with the dolphins for the same fish species, the tambaqui (*Colossoma macropomum*) and the pirapitinga (*Piaractus brachypomus*), resulting in deliberate or unintentional catches of the Amazon river dolphin. The local fishermen overfish and when the Amazon River dolphins remove the commercial catch from the nets and lines, it causes damages to the equipment and the capture, as well as generating ill will from the local fishermen. The negative reactions of the local fishermen are also attributed to the Brazilian Institute of Environment and Renewable Natural Resources prohibition on killing the Amazon river dolphin, yet not compensating the fishermen for the damage done to their equipment and catch.

During the process of catching the commercialized fish, the Amazon river dolphins get caught in the nets and exhaust themselves until they die, or the local fishermen deliberately kill the entangled dolphins. The carcasses are discarded, consumed, or used as bait to attract a scavenger catfish, the piracatinga (*Calophysus macropterus*). The use of the Amazon river dolphin carcass as bait for the piracatinga dates back to 2000. Increasing demand for the piracatinga has created a market for distribution of the Amazon river dolphin carcasses to be used as bait throughout these regions.

Of the 15 dolphin carcasses found in the Japurá River in 2010–2011 surveys, 73% of the dolphins were killed for bait, disposed of, or abandoned in entangled gillnets. The data do not fully represent the actual overall number of deaths of the Amazon river dolphins, whether accidental or intentional, because a variety of factors make it extremely complicated to record and medically examine all the carcasses. Scavenger species feed upon the carcasses, and the complexity of the river currents make it nearly impossible to locate all of the dead animals. More importantly, the local fishermen do not report these deaths out of fear that a legal course of action will be taken against them, as the Amazon river dolphin and other cetaceans are protected under a Brazilian federal law prohibiting any takes, harassments, and kills of the species.

Conservation

In 2008, the International Whaling Commission (IWC) expressed concern for captured botos for use as bait in the Central Amazon, which is an emerging problem that has spread on a large scale. The species is listed in Appendix II of the Convention on International Trade in Endangered

Species Fauna and Flora (CITES), and Appendix II of the Convention on the Conservation of Migratory Species of Wild Animals, because it has an unfavorable conservation status or would benefit significantly from international co-operation organized by tailored agreements.

According to a previous assessment by the Scientific Committee of the International Whaling Commission in 2000, the population of botos appears great and there is little or no evidence of population decline in numbers and range. However, increased human intervention on their habitat is expected to, in the future, be the most likely cause of the decline of its range and population. A series of recommendations were issued to ensure proper follow-up to the species, among which is the implementation and publication of studies on the structure of populations, making a record of the distribution of the species, information about potential threats as the magnitude of fishing operations and location of pipelines.

In September 2012, Bolivian President Evo Morales enacted a law to protect the dolphin and declared it a national treasure.

In 2008, the species was listed on the Red list of endangered species, but in 2011, the IUCN stated it as Data Deficient (DD). The species was previously listed as "vulnerable" but the conservation status changed due to the limited amount of currently available information on threats, ecology, and population trends. In areas where these dolphins have been studied, they appear well extended and relatively abundant. However, these areas represent only a small proportion of the total distribution of the species and are often sites where the animals are protected. Consequently, the information from these areas may not be representative, and may not be valid in the long term.

Increasing pollution and gradual destruction of the Amazon rainforest add to the vulnerability of the species. The biggest threats are deforestation and other human activities that contribute to disrupt and alter their environment. Another source of concern is the difficulty in keeping these animals alive in captivity, due to intra-species aggression and low longevity. Captive breeding is not considered a conservation option for this species.

In Mythology

In traditional Amazon River folklore, at night, an Amazon river dolphin becomes a handsome young man who seduces girls, impregnates them, and then returns to the river in the morning to become a dolphin again. Similarly, the female becomes a beautiful, well - dressed, wealthy - looking and young woman. She goes to the house of a married man, places him under a spell to keep him quiet, and takes him to a thatched hut and visits him every year on the same night she seduced him. On the 7th night of visiting, she changes the man into a female - baby or male, and soon transfers it into his own wife's womb. The mythology is said to be the cycle of a baby. This dolphin shape shifter is called an *encantado*. The myth has been suggested to have arisen partly because dolphin genitalia bear a resemblance to those of humans. Others believe the myth served (and still serves) as a way of hiding the incestuous relations which are quite common in some small, isolated communities along the river. In the area, tales relate it is bad luck to kill a dolphin. Legend also states that if a person makes eye contact with an Amazon river dolphin, he or she will have lifelong nightmares. Local legends also state that the dolphin is the guardian of the Amazonian manatee, and that if one should wish to find a manatee, one must first make peace with the dolphin.

Associated with these legends is the use of various fetishes, such as dried eyeballs and genitalia. These may or may not be accompanied by the intervention of a shaman. A recent study has shown, despite the claim of the seller and the belief of the buyers, none of these fetishes is derived from the *boto*. They are derived from *Sotalia guianensis*, are most likely harvested along the coast and the Amazon River delta, and then are traded up the Amazon River. In inland cities far from the coast, many, if not most, of the fetishes are derived from domestic animals such as sheep and pigs.

Southern Elephant Seal

The southern elephant seal (Mirounga leonina) is the world's largest seal and the largest member of the order Carnivora, which includes all seals, cats, dogs, foxes, wolves, bears, raccoons, hyenas, weasels and civets, among others. Southern elephant seals are "true"" or "earless" seals and are named for the male's trunk-like proboscis (nose) and for the seal's massive size. The species was hunted almost to extinction in the 19th and 20th centuries and, on South Georgia, their blubber was rendered into oil as late as 1964. Since that time, the southern elephant seal population has recovered to about 700,000 individuals.

Distribution and Migration of Southern Elephant Seals

The southern elephant seal spends most of the year at sea (9–10 months), migrating anywhere from sub-Antarctic waters to nearly as far north as the Tropic of Capricorn in search of food. Seasonal distribution is poorly known, but males are believed to forage in the waters close to the Antarctic continental shelf while females circulate farther into more distant open waters.

During the breeding season (August–November), individuals go ashore on the sub-Antarctic islands. Males arrive in late August to claim and fight for territory, and females arrive late September through early October, giving birth just after arrival. Southern elephant seals also go ashore and fast during the molting season, which lasts three to five weeks during January and February.

Adult male southern elephant seal at Royal Bay

South Georgia is home to the largest population of southern elephant seals, hosting more than half of the world's population. Other notable populations are found at Macquarie Island, Heard Island, the Falkland and the Kerguelen Islands, and at widespread islands near the Antarctic Peninsula.

Elephant seals return annually to the same place for breeding, birthing and molting, and probably to the same hunting areas of the open ocean.

Identifying Southern Elephant Seals

Physical Characteristics

Southern elephant seals are enormous the largest of all seals. Males can be over 20 feet (6m) long and weigh up to 8,800 pounds (4,000kg) or more. A species with extreme sexual dimorphism, males are typically five to six times larger than the females, who weigh 880–2,000 pounds (400–900kg). Males are fully grown by their eighth year.

Southern elephant seals have large blocky heads, strong front flippers used to steer while they swim, and rear flippers used for propulsion. Complementing their tremendous bulk is the inflatable trunk-like proboscis of the bull. The southern elephant seal's large eyes indicate that vision is important for hunting in the dark ocean depths.

When onshore, seals of all sexes and ages display a snotty-looking discharge from their noses (much to the chagrin of many photographers). This discharge is actually the method by which excess salt, from ingested seawater, is eliminated from their bodies. To avoid predation, young southern elephant seals have darker backs and lighter bellies. This allows camouflage by blending in with the lighter water when viewed from below and with the darker water when seen from above. Orcas and great white sharks are their principal predators mostly preying on inexperienced young at sea.

Pups are born with dark brown-black fur that is not adapted for extended periods in the cold water. Most of the time pups stay on land in their black "lanugo" pelage, unless they are forced into the water. After weaning, the pup's coat molts and the "weaner" starts to learn to swim and hunt. Young "weaners" are often mistaken by inexperienced nature travelers for other species of seals mostly Weddell seals. Adult skin, after molting, is silvery-brown but on land it is usually browner from staining by the mud and excrement that fills the "wallows" in which the seals often congregate while on the shore.

Bull elephant seal bellowing

Southern elephant seals are closely related to northern elephant seals that look remarkably similar, although the northern is noticeably smaller. Northern elephant seals are found in the eastern

Pacific along the west coast of North America from the Baja Peninsula to Alaska. Their range does not overlap with southern elephant seals.

Southern Elephant Seal Habitat

When going ashore to breed or molt, the southern elephant seal seeks rocky island shores and smooth sandy beaches. The rest of the year is spent entirely at sea replenishing their layer of blubber.

Southern Elephant Seal Breeding and Mating

Southern elephant seals give birth and breed September to November. The larger males arrive at the breeding beach a month before the females and the smaller/younger males in order to fight for dominance and the right to breed with a harem of females.

On shore, the males start to form beach colonies. The strongest bulls the alpha males establish harems of several dozen females, with the biggest males occupying the beach real estate most desirable for attracting females. Beta males have smaller harems. The least successful males have no harems, but may try to seduce another male's females. An elephant seal must stay in his territory to defend it, which can mean fasting for up to three months, living off stored blubber. During this time the male can lose as much as 40% of his body weight. As the mating season progresses the biggest dominant bulls can acquire a harem of up to 50 females or more. Occasionally, the number of breeding females in a harem can exceed 100 seals.

The amount of time spent ashore by males during the breeding season varies greatly; some breeding males spend more than 60–90 days on shore—the females spend 25–30 days.

The female gives birth 0–10 days after coming ashore and does not leave the beach to feed until her pup is weaned. During this time she depends on her stored fat reserves to sustain her and loses an average of 35% of her body weight, a weight loss of 17 lbs (8kg) per day.

Pups weigh about 88 pounds (40kg) at birth and reach 260–290 pounds (120–130kg) by the time they are weaned. The nursing period lasts an average of about 23 days and the pup puts on weight very quickly during this period as the mother's milk contains as much as 52% fat. Pups can gain 7–9 pounds (3.5–4kg) per day. The mother mates up to 5 days before her pup is weaned and upon weaning she abandons her pup and returns to the sea. By this time, the pups have quadrupled their weight.

Male and female southern elephant seals

The pup leaves the beach about 3–8 weeks later to begin the process of learning to find food. Suitable food may not be located near its birth beach, so the pup relies on its stored body fat to get it through this lean time—an important aspect for its survival. The weaned size of pups is variable and some are three times the weight of others.

Many pups die on the crowded breeding beaches—some crushed by the giant males, others abandoned by their mothers.

Most females reach sexual maturity at 2–4 years of age. Males may reach sexual maturity at 3–6 years of age but few breed until 10 years of age.

Southern Elephant Seal Diet

Southern elephant seals feed exclusively at sea. The main sources of food are fish and squid. Prey is captured both near the surface and very deep underwater. The seals have been known to eat bottom dwelling fish from the ocean's darkest depths. Southern elephant seals locate prey with their vision; their large eyes are an adaptation allowing them to take advantage of the bioluminescence of some prey. While they do not use echolocation, it is believed that their vibration-sensitive whiskers also aid them in locating prey.

Southern Elephant Seal Social Behavior

During the foraging months, southern elephant seals spend 90% of their time underwater, surfacing approximately every 20–30 minutes for a short rest of 2–3 minutes. They are swift and powerful swimmers and can dive further and longer than any other seal. A physiological adaptation allows them to store oxygen and simultaneously reduce its consumption while diving. They can lower their heart rate to as little as a single beat per minute. The recorded diving record for the seal is nearly two hours underwater and more than 4,200 feet (1,400m) in depth. At sea, they are usually solitary animals.

During breeding and molting, they are more social, gathering in large colonies along their natal beaches. Males defend their breeding territories through a series of specific behaviors. An interloper will vocally (their large proboscis amplifies their bellows) and physically challenge the alpha male, who in turn will rise up on its chest and respond in kind. If this does not drive away the challenger, vicious and bloody fights can break out, with the winner keeping or claiming the territory and local females as his own. Subordinate males will exhibit a flattened posture without inflating their proboscis when near another male's harem to demonstrate that they are not a threat.

Colonies form again in January and February when the southern elephant seals go through a molting process of shedding their fur and the outer layer of skin. After three to five weeks, they grow new fur and return to the sea. Southern elephant seals are known to live up to 23 years.

Walrus

The walrus (Odobenus rosmarus) is a large Arctic marine mammal with flippers, a broad head, short muzzle, small eyes, tusks and whiskers. Scientists recognize two subspecies of walrus – the Atlantic walrus and the Pacific walrus.

Walruses are cinnamon brown in color. They are able to turn their hind flippers forward to aid in movement on land. Their front flippers are large and each has five digits. Males have special air sacs that are used to make a bell-like sound.

The mustached and long-tusked walrus is most often found near the Arctic Circle, lying on the ice with hundreds of companions. These marine mammals are extremely sociable, prone to loudly bellowing and snorting at one another, but are aggressive during mating season. With wrinkled brown and pink hides, walruses are distinguished by their long white tusks, grizzly whiskers, flat flipper, and bodies full of blubber.

Walrus Tusks

Walruses use their iconic long tusks for a variety of reasons, each of which makes their lives in the Arctic a bit easier. They use them to haul their enormous bodies out of frigid waters, thus their "tooth-walking" label, and to break breathing holes into ice from below. Their tusks, which are found on both males and females, can extend to about three feet, and are, in fact, large canine teeth, which grow throughout their lives. Male walruses, or bulls, also employ their tusks aggressively to maintain territory and, during mating season, to protect their harems of females, or cows.

Arctic Adaptations

The walrus' other characteristic features are equally useful. As their favorite meals, particularly shellfish are found near the dark ocean floor, walruses use their extremely sensitive whiskers, called mustacial vibrissae, as detection devices. Their blubbery bodies allow them to live comfortably in the Arctic region—walruses are capable of slowing their heartbeats in order to withstand the polar temperatures of the surrounding waters.

Atlantic and Pacific Subspecies

The two subspecies of walrus are divided geographically. Atlantic walruses inhabit coastal areas from northeastern Canada to Greenland, while Pacific walruses inhabit the northern seas off Russia and Alaska, migrating seasonally from their southern range in the Bering Sea—where they are found on the pack ice in winter—to the Chukchi Sea. Female Pacific walruses give birth to calves during the spring migration north.

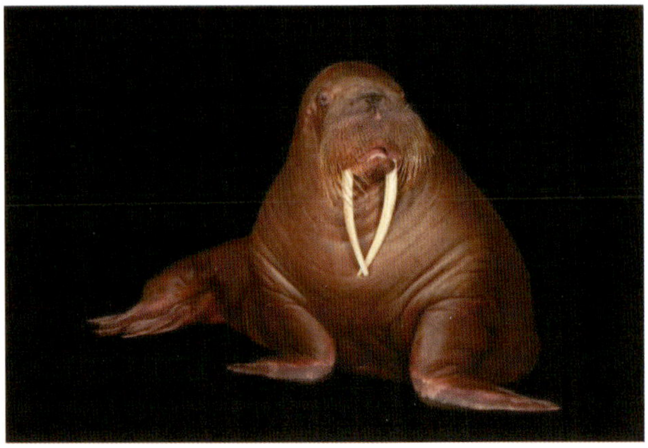

Both males and females have large tusks that are used for defense, cutting through ice and getting out of the water. The tusks can be more than three feet long in males and about two and a half feet long in females.

Diet

Walrus staples include clams, mussels and other bottom dwelling (or benthic) organisms that they locate through their whiskers. They are also known to eat carcasses of young seals when food is scarce.

Population

The worldwide walrus population is about 250,000 animals. Pacific walruses number more than 200,000. The Pacific walrus population was severely reduced by hunting in the past, but their numbers have rebounded after these severe reductions.

Range

The walrus is circumpolar in its range but they are found in geographically separate areas. The Pacific walrus is found in the Bering, Chukchi, and Laptev Sea, while the Atlantic walrus inhabits the coastal regions of northeastern Canada and Greenland.

Behavior

Walruses are very social animals and congregate in large numbers. They haul out in herds and males and females form separate herds during the non-breeding season. They establish dominance through threat displays involving tusks, bodies and aggression. The largest walruses are the most aggressive. Walruses spend two thirds of their lives in and out of the water, feeding and resting on sea ice. Most walrus groups migrate north in the summer and south in the winter, and females haul out on the ice to give birth.

Reproduction

Calves are ashen gray to brown in color and weigh in from about 99-165 pounds at birth. They turn reddish brown within a few weeks and grow rapidly on their mothers' milk.

- Mating Season: Between December and March.
- Gestation: 15-16 months.
- Offspring: Generally one calf, though twins have been recorded.

References

- Zofia Kielan-Jaworowska, Richard L. Cifelli, Zhe-Xi Luo (2004). “Chapter 7: Eutriconodontans”. Mammals from the Age of Dinosaurs: origins, evolution, and structure. New York: Columbia University Press. pp. 216–248. ISBN 0-231-11918-6
- Meng, Jin; Hu, Yaoming; Wang, Yuanqing; Wang, Xiaolin; Li, Chuankui (2007). “Corrigendum: A Mesozoic gliding mammal from northeastern China”. Nature. 446 (7131): 889–893. Bibcode:2007Natur.446Q.102M. doi:10.1038/nature05639
- Aquatic-mammals-and-adaptations-chordata-zoology-8523: notesonzoology.com, Retrieved 15 March 2018
- Haw, Jim; Handy, Dana (2013). «Northern Elephant Seals: Increasing Population, Decreasing Biodiversity». Scientific American. Retrieved 28 June 2016
- Bruno, Díaz L. (2006). “Interactions between Mediterranean bottlenose dolphins (Tursiops truncatus) and gillnets off Sardinia, Italy”. Journal of Marine Science. 63 (5): 946–951. doi:10.1016/j.icesjms.2005.06.012
- What-is-a-marine-mammal: whalefacts.org, Retrieved 25 May 2018
- Kielan-Jaworowska, Zofia; Cifelli, Richard L.; Luo, Zhe-Xi (2004). Mammals from the Age of Dinosaurs: Origins, Evolution, and Structure. New York: Columbia University Press. pp. 441–462. ISBN 978-0-231-50927-5
- Gaetano, Leandro C.; Rougier, Guillermo W. (2011). “New materials of Argentoconodon fariasorum (Mammaliaformes, Triconodontidae) from the Jurassic of Argentina and its bearing on triconodont phylogeny”. Journal of Vertebrate Paleontology. 31 (4): 829–843. doi:10.1080/02724634.2011.589877
- Amazon-river-dolphin, species: worldwildlife.org, Retrieved 29 March 2018
- «Turbidity: Description, Impact on Water Quality, Sources, Measures» (PDF). Minnesota Pollution Control Agency. March 2008. Retrieved 16 June 2016
- Harwood, J. (2001). “Marine Mammals and their Environment in the Twenty-First Century”. Journal of Mammalogy. 82 (3): 630–640. doi:10.1644/1545-1542(2001)082<0630:MMATEI>2.0.CO;2. JSTOR 1383602
- Southern-elephant-seal, antarctica-wildlife: travelwild.com, Retrieved 15 May 2018
- Jefferson, T. A.; Webber, M. A.; Pitman, R. L. (2009). Marine Mammals of the World A Comprehensive Guide to their Identification (1st ed.). London: Academic Press. pp. 7–16. ISBN 978-0-12-383853-7. OCLC 326418543

Chapter 5

Primates

A primate is a mammal that belongs to the order Primates. This chapter discusses in extensive detail about primates and in particular, the eastern gorilla, which is the largest species within the order.

Gorilla

Gorillas are the largest members of the primate family and are closely related to humans, with 98% of their DNA identical to that of humans. Unlike other primates they are terrestrial, meaning they do not climb trees and are land dwelling, inhabiting the tropical rainforests of central Africa. Historically the gorilla has been portrayed as a vicious killer; however they are shy gentle creatures that would not attack humans unless provoked.

Gorillas are classified as mammals, which is defined by the Oxford English dictionary as a 'Warm-blooded vertebrate animal that has hair or fur, secretes milk, and (typically) bears live young'. Upon their discovery gorillas were classified as one species, however they are now separated into two species and four sub species according to geographical location and physical characteristics. They live in groups called a troop lead by a dominant silverback male, and are highly sociable animals, maintaining strong bonds between group members.

In the wild their only predators are leopards, which can attack vulnerable youngsters and the biggest threat to gorillas is man. In recent decades gorilla populations have been affected by habitat loss, disease and poaching. Subsequently all gorilla species are classified as endangered by the International Union for Conservation of Nature (IUCN). Due to the vast dense areas in which they live it is difficult to monitor and protect gorilla populations and therefore difficult to implement successful conservation techniques. However conservation efforts persist and several governmental and non governmental organizations named 2009 the 'Year of the Gorilla' with the aim of working together to further improve gorilla conservation.

Species Identification

There are two recognized gorilla species the Eastern Gorilla and Western Gorilla, defined by the areas in Africa which they inhabit. Within these classifications are four subspecies including the Eastern Lowland Gorilla, Mountain Gorilla, Cross River Gorilla and Western Lowland Gorilla. All subspecies live in varying habitats, have unique physical characteristics and have differing populations.

Eastern Gorilla (Gorilla beringei)

Two subspecies exists:

- Species: Eastern Lowland Gorilla
- Scientific name: Gorilla beringei graueri
- Location: Democratic Republic of Congo
- Population: Less than 3000
- Conservation status: Endangered
- Physical characteristics: Largest of gorilla sub species, longer arms than the mountain gorilla and shorter hair and teeth.
- Species: Mountain Gorilla
- Scientific name: Gorilla beringei beringei
- Population: Less than 720
- Conservation Status: Critically Endangered
- Physical characteristics: Large skull, wide face and angular nostrils. Larger body and longer hair than eastern lowland gorilla.

Western Gorilla (Gorilla Gorilla)

Two subspecies exists:

- Species: Western Lowland Gorilla
- Scientific name: Gorilla gorilla gorilla
- Location: Democratic Republic of Congo
- Population: 100,000
- Conservation status: Critically Endangered
- Physical characteristics: Males silverback coloring extends onto the thighs, also have redder hair on their heads.
- Species: Cross River Gorilla
- Scientific name: Gorilla gorilla diehli
- Population: Approximately 300
- Conservation Status: Critically Endangered
- Physical characteristics: Smaller skull and teeth than other gorilla as well as shorter hands and feet.

Physical Characteristics

Gorillas are quadrupedal meaning they move around by walking on their knuckles, their huge arms provide balance and support for their large frames. Size and weight varies between species and gender, with males reaching heights of 1.7 m when stood tall and weighing between 170 kg to 240 kg. Females usually weigh between 70 kg to 100 kg and are shorter reaching heights of 1.5 m.

Physical characteristics of gorilla

Evolution

Several million years ago the gorilla split from its closest living relative the chimpanzee. It is thought that originally gorilla were just one species, but changed as they spread and adapted to different living conditions.

As gorillas are ground dwelling they have evolved without a tail, which is used by other primates for balancing and gripping when climbing trees. They have good hearing, which is necessary in thick vegetation where visibility can be poor and good eyesight which they use to identify food.

Social Structure

Gorilla's live in groups called a troop, which can be made up of 2- 40 animals. Within this group there is one dominant silverback male several females and younger males and offspring of different ages.

Silverbacks

Are large adult males with distinctive silver coloured hair on their back. This is developed through maturity and is usually present on male Gorilla's over 12 years of age. An adult silverback can weigh over 180 kg and is the biggest and heaviest of all living primates. As the leader of the group the dominant silverback is in charge of protecting its troop from predators or rival silverbacks, displaying a broad chest and large canine teeth when threatened. The dominant silverback is also in charge of finding good places to forage for food and safe places to rest and sleep. If the dominant silverback dies then a troop will disband and form new groups.

Blackbacks

Younger male gorillas are called blackbacks because they have not yet developed silver hair on their back like mature males, though all males will not necessarily develop a silver back. They support the leader and help them forage. When they reach sexual maturity at the age of ten they leave their family groups to join another or form their own and find a mate. They usually travel alone or with a group of other males until they can attract a mate.

Females

Females leave their family groups when they reach sexual maturity in order to prevent in breeding. They can join lone male or other females and within groups are ranked according to when they joined the group; this rank is passed on to their offspring. They form strong relationships with males through behavior such as grooming, which is important as male gorillas provide protection from predators.

Juveniles and Infants

Young infants are playful and inquisitive, and for the first 3-4 years of their lives they have close bond with their mother. Juveniles will groom each other and the dominant silverback will groom juveniles and protects them.

Relationships

Gorillas are social animals and within troops exist different relationships. All females in the troop have a strong relationship with the dominant silverback, which is strengthened through bonding activities such as grooming. However males and females are often aggressive to each other, though it rarely results in death.

Female to female relationships are not strong and often aggressive fighting occurs over access to males. Relationships between males are not strong because of dominance issues however within all male groups individuals are friendlier with each other.

Gorillas are tolerant of other species and would only attack if they felt threatened. They are wary of humans and it takes a long time to build their trust.

Daily Activities

Gorillas walk around 1 km each day at a slow pace foraging for food. They wake up early between 6-8 am and spend the day eating and playing and relaxing. Each evening between 5-6 pm they build a bowl shaped nest from leaves and vegetation to sleep in, which keeps them warm and only a mother and her offspring share a nest.

Intelligence

Language

Some gorillas in captivity have been taught to communicate with humans. As part of a gorilla language project, Koko a western lowland gorilla has been taught American Sign Language and has mastered 1,000 words.

Use of Tools

Younger male gorillas are called blackbacks because they have not yet developed silver hair on their back like mature males, though all males will not necessarily develop a silver back. They support the leader and help them forage. When they reach sexual maturity at the age of ten they leave their family groups to join another or form their own and find a mate. They usually travel alone or with a group of other males until they can attract a mate.

Communication

- Vocal

  Gorillas produce over 20 recognized sounds, each with unique meaning. Unlike humans they have large Laryngeal sacs within their vocal chamber which can be filled with up to seven litres of air, and used to increase the volume of their calls.

- Visual

  Gorillas, like humans use a series of facial expressions and body movements to convey messages and emotions.

- Aggression

  Chest beating, staring with pursed lips and a tensed mouth with teeth bared.

- Fear

  Open mouth grimace with raised eyebrows and tilted head.

- Distress

  Pursed lips and raised eyebrows, used by infants when left alone.

- Seduction

  A strutting walk and bristled hair occasionally glancing at the female. This can be used as an attempt to lure females away from groups.

Olfactory

The silverback releases chemicals called pheromones from glands in his armpits, which have a strong odor. This is known as the 'fear smell' as it can warn other gorilla of impending danger and is a signal of excitement and aggression.

Tactile

Tactile communication is very important for gorillas as it helps them to create and maintain strong social bonds. They touch each others face and hands for reassurance and groom each other regularly. Within groups lower ranking members usually groom high ranking members and adult females do not groom each other.

Reproduction and Care of Young

Gorillas have very slow rates of reproduction and high rates of infant mortality, as many as 40% do not survive after birth. They live in polygamous groups, meaning the dominant male mates with all the females and mating occurs when the female is mid way through her cycle. Infanticide (the killing of young) takes place on occasions when a dominant silverback dies and a new male replaces him, killing all his offspring. Mothers are very protective of their young and maintain a strong bond with them through close physical contact; the dominant silverback will also interact with young, though primary care is left to the mother.

Female gorillas in the wild become sexually mature at around 10 years old. However, males mature later and don't usually reproduce before 15 years old. Mating activity is usually initiated by the female and occurs all year round. Gestation takes approximately 8.5 months and a single infant is produced. Baby gorillas learn to crawl at around two months, can ride on their mothers backs at seven months and walk at nine months. They are breast fed for the first three years of their life, but begin to eat vegetation at around six months.

Diet

Gorillas are herbivores and therefore only eat fruit and leaves. They do not hunt or eat other animals, but occasionally feed on small insects and termites. Diet varies between species mainly due to the availability of certain foods in different habitats. Some gorilla species can consume over 200 species of plants and fruit.

Species	Food eaten
Mountain Gorilla	Nettles, bamboo and celery. Particularly enjoy eating wild berries when available. Consume over 142 species of plant. One percent of its diet composed of ants and other insects.
Western Lowland Gorilla	Prefer sugary fruit, seeds and stems. Usually eat more fruit in the wet season and herbs and bark in the dry season.
Eastern Lowland Gorilla	Mainly leaves and vegetation. Eat some fruits according to seasonality.
Cross River Gorilla	More tree bark and leaves. Eat less fruit.

Diseases

Recent research has shown that the parasite Plasmodium falciparum that aggressively infects millions of humans each year has been found in gorillas, along with two new species of malaria parasites (Plasmodium GorA and Plasmodium GorB). Plasmodium falciparum originated from a closely related parasite found in chimpanzees in equatorial Africa and is responsible for 85 percent of malaria infections in humans and nearly all deaths from the disease. Increase in logging, deforestation, and bushmeat hunting favors the transmission of new parasites to humans from gorillas and vice versa.

Ebola virus is extremely lethal for humans and other great apes including gorillas. Between 2001 and 2005, at least 5,500 were killed at the Lossi Sanctuary in northwest Republic of Congo. Recorded outbreaks of the Ebola virus since 2000 might have claimed thousands of gorillas in Africa. It is unclear how the disease is spread but likely through the butchering and handling of primate bushmeat. Mountain gorillas have not yet been affected. The conservation implications of Ebola virus are significant since past outbreaks have occurred mainly in remote protected areas that were the stronghold for gorilla protection.

Threats

Bushmeat Trade

In some parts of Africa gorillas are hunted for their meat. Animals are caught using snares or riffles and taken to markets usually far away. Sometimes gorillas are unintentionally caught in snares designed to trap other animals. Though such trade is illegal, people go to great lengths to avoid detection and trade can be difficult to monitor. In particular western species of gorilla have been most affected by the bushmeat trade. A recent undercover investigation conducted by Endangered Species International has found that up to two gorillas are killed and sold as bushmeat each week a region of the Republic of Congo.

Habitat Loss

Destruction of forest areas due to commercial logging and agriculture is having a huge effect on gorilla populations. Approximately 80% of gorillas live in unprotected areas and are therefore exposed to human influence on their environment. The opening up of new areas for industries has brought a new labor force into remote areas, where bushmeat is hunted for sustenance. Improved infrastructure such as the building of roads has also given illegal trappers easier access to gorillas and caused habitat destruction. Accidental or uncontrolled bushfires can also spread through habitats stripping areas of vegetation that gorillas could feed on. Due to their shy natures and wariness of humans gorillas are pushed into smaller areas away from human activity.

Conflict/Civil War

In the past war has forced refugees into areas of gorilla habitat and with limited access to food people are pressured to hunt for bushmeat. Human presence in areas of gorilla habitat also increases the risk of human to gorilla disease transmission. With political instability and limited funding monitoring and protecting gorilla populations becomes difficult.

Disease

Gorillas are susceptible to the same diseases as humans and are particularly vulnerable to respiratory infections. Any form of contact with humans or human debris can transmit diseases to gorillas. As a preventative measure tourists have to abide by strict regulations when viewing gorillas keeping a safe distance away. The Ebola virus is another huge problem facing gorillas as it spreads quickly and in 80% of cases leads to death. Scientist Peter Walsh stated 'A quarter of the gorillas in the world have died form Ebola in the last 12 years' and found in one study area that 5000 gorillas had died from the disease.

Pets/Souvenirs

Some gorillas are hunted for their body parts for medicinal use and hands, skulls and feet are sold illegally as souvenirs. In some cases infants are stolen to be sold as pets, in these situations adults are usually killed as they will fight to the death to protect their young.

Conservation

All species of gorilla are listed by the International Union for Conservation of Nature and Natural Resources (IUCN) as endangered with the Cross River Gorilla and Mountain Gorilla listed as critically endangered, with some populations showing a decreasing population trends. Several measures have been taken by governments and non governmental organizations to combat the major threats of habitat loss, disease and poaching. Below are some of the steps that are being taken to prevent further population decrease.

Education is an important tool for making people aware of the importance of conserving gorillas. Many organizations like Endangered Species International run outreach programmes teaching people ways to protect gorillas, their value and ways to improve hygiene to stop the spread of disease.

Combat disease with the development of vaccinations. At present scientist are developing a vaccination against Ebola however this would be very expensive to implement.

Increased protection is needed to protect gorillas from poachers. An increase in the number of forest rangers, areas of protection and better monitoring of populations is much needed to prevent further deaths from hunting and disease.

Eco Tourism can provide money which can be directed towards environmental initiatives and provide locals with alternative sources of income. However in the case of gorillas it must be strictly controlled to prevent the spread of disease and disturbance to the animals.

Eastern Gorilla

The eastern gorilla is one of two gorilla sub-groups found roaming in the jungles on the African continent (the other being the western gorilla). The eastern gorilla is less common than the western gorilla and is also the largest primate in the world.

The eastern gorilla is found inhabiting the tropical jungles and forests of parts of eastern and central Africa, along with lowland swamps and secondary forests. The eastern gorilla is found in the trees in throughout the countries of Rwanda, Uganda and eastern Congo.

There are considered to be two separate sub-species of eastern gorilla which are the eastern lowland gorilla and the mountain gorilla. Although only slightly different in appearance, the two eastern gorilla species differ in their numbers with the mountain gorilla being rarer in the wild today than the eastern lowland gorilla.

The eastern gorilla is one of the great apes, a group that includes orangutans, gorillas, humans and chimpanzees. As with the other great apes, the eastern gorilla has a number of features which makes living in the jungle a bit easier, including having opposable thumbs which come in handy when the eastern gorilla is peeling fruit.

The eastern gorilla is an omnivorous animal, but the majority of its diet is made up of eating fruit which the eastern gorilla is known to travel vast distances through the forests to find. The eastern gorilla also eats leaves, nuts and berries, along with insects and occasionally small animals such as lizards and rodents. The eastern gorilla has also been observed using basic tools in the wild in order to more effectively gather food.

Due to its large size, the eastern gorilla has few real predators in its native African forests, with large cats such as leopards and the odd crocodile being the only real natural threat to the eastern gorilla. The biggest threat to the eastern gorilla is habitat loss caused by deforestation and also being hunted by humans. Parts of the eastern gorilla's territory has also been taken over by civil unrest in recent years, which, along with poaching, has had a truly devastating affect on wild populations.

The eastern gorilla tends to live in groups which are led and protected by the alpha male. The alpha male eastern gorilla also mates with the females in his group, producing generally single offspring, known as babies. The eastern gorilla babies remain with their mother until they are a few years old and become independent.

Today, all eastern gorillas are considered to be critically endangered species with populations having suffered due to habitat loss and illegal poaching. There are estimated to be around 5,000 eastern lowland gorillas, and an even fewer 700 mountain gorilla individuals left in the wild.

References

- Murray E. Fowler; Susan K. Mikota (2006). Biology, Medicine, and Surgery of Elephants. John Wiley & Sons. p. 353. ISBN 978-0-8138-0676-1
- Payne, K.B.; Langbauer, W.R.; Thomas, E.M. (1986). "Infrasonic calls of the Asian elephant (Elephas maximus)". Behavioral Ecology and Sociobiology. 18(4): 297–301. doi:10.1007/BF00300007
- African-bush-elephant: animalsadda.com, Retrieved 28 March 2018
- Bekoff, M. (2009). "Anthropomorphic Double-Talk: Can Animals Be Happy But Not Unhappy? No!". Retrieved 5 September 2013
- Martin, F.; Niemitz C. (2003). ""Right-trunkers" and "left-trunkers": side preferences of trunk movements in wild Asian elephants (Elephas maximus)". Journal of Comparative Psychology. 117(4): 371–79. doi:10.1037/0735-7036.117.4.371. PMID 14717638
- Loxodonta, Africana, accounts: animaldiversity.org, Retrieved 08 July 2018
- Estes, R. (1991). The behavior guide to African mammals: including hoofed mammals, carnivores, primates. University of California Press. pp. 263–66. ISBN 978-0-520-08085-0
- Plotnik, J. M.; de Waal & F. B. M.; Reiss, D. (2006). "Self-recognition in an Asian elephant". Proceedings of the National Academy of Sciences. 103 (45): 17053–57. Bibcode:2006PNAS..10317053P. doi:10.1073/pnas.0608062103. PMC 1636577
- African-bush-elephant: animalsadda.com, Retrieved 22 April 2018
- Blanc, J. (2008). "Loxodonta africana". IUCN Red List of Threatened Species. Version 2012.2. International Union for Conservation of Nature. Retrieved 2010-04-04
- Jachmann, H.; Berry, P. S. M.; Imae, H. (1995). "Tusklessness in African elephants: a future trend". African Journal of Ecology. 33 (3): 230–235. doi:10.1111/j.1365-2028.1995.tb00800.x

Permissions

All chapters in this book are published with permission under the Creative Commons Attribution Share Alike License or equivalent. Every chapter published in this book has been scrutinized by our experts. Their significance has been extensively debated. The topics covered herein carry significant information for a comprehensive understanding. They may even be implemented as practical applications or may be referred to as a beginning point for further studies.

We would like to thank the editorial team for lending their expertise to make the book truly unique. They have played a crucial role in the development of this book. Without their invaluable contributions this book wouldn't have been possible. They have made vital efforts to compile up to date information on the varied aspects of this subject to make this book a valuable addition to the collection of many professionals and students.

This book was conceptualized with the vision of imparting up-to-date and integrated information in this field. To ensure the same, a matchless editorial board was set up. Every individual on the board went through rigorous rounds of assessment to prove their worth. After which they invested a large part of their time researching and compiling the most relevant data for our readers.

The editorial board has been involved in producing this book since its inception. They have spent rigorous hours researching and exploring the diverse topics which have resulted in the successful publishing of this book. They have passed on their knowledge of decades through this book. To expedite this challenging task, the publisher supported the team at every step. A small team of assistant editors was also appointed to further simplify the editing procedure and attain best results for the readers.

Apart from the editorial board, the designing team has also invested a significant amount of their time in understanding the subject and creating the most relevant covers. They scrutinized every image to scout for the most suitable representation of the subject and create an appropriate cover for the book.

The publishing team has been an ardent support to the editorial, designing and production team. Their endless efforts to recruit the best for this project, has resulted in the accomplishment of this book. They are a veteran in the field of academics and their pool of knowledge is as vast as their experience in printing. Their expertise and guidance has proved useful at every step. Their uncompromising quality standards have made this book an exceptional effort. Their encouragement from time to time has been an inspiration for everyone.

The publisher and the editorial board hope that this book will prove to be a valuable piece of knowledge for students, practitioners and scholars across the globe.

Index